SPACE-TIME VARIATION OF HYDROLOGICAL PROCESSES AND WATER RESOURCES IN RWANDA
FOCUS ON THE MIGINA CATCHMENT

Space-time variation of hydrological processes and water resources in Rwanda
Focus on the Migina catchment

DISSERTATION

Submitted in fulfillment of the requirements of
the Board for Doctorates of Delft University of Technology
and of the Academic Board of the UNESCO-IHE Institute for Water Education
for the Degree of DOCTOR
to be defended in public
on Wednesday, 7 May 2014 at 10:00 hours
in Delft, the Netherlands

by

Omar MUNYANEZA

Master of Science in Water Resources and Environmental Management,
National University of Rwanda (NUR), Butare, Rwanda

born in Umutara, Gatsibo District, Rwanda

This dissertation has been approved by the promoter:
Prof. dr. S. Uhlenbrook

Composition of the Doctoral Committee:

Chairman	Rector Magnificus TU Delft
Vice-chairman	Rector UNESCO-IHE
Prof. dr. S. Uhlenbrook	UNESCO-IHE/Delft University of Technology, Promoter
Prof. dr. V.G. Jetten	Twente University/ITC
Prof. K.A. Irvine	UNESCO-IHE/Wageningen University
Prof. dr. ir. H.H.G. Savenije	Delft University of Technology/UNESCO-IHE
Prof. dr. ir. U.G. Wali	University of Rwanda, Butare, Rwanda
Dr. S. Maskey	UNESCO-IHE
Prof. dr. ir. N.C. van de Giesen	Delft University of Technology (reserve member)

The research reported in this dissertation has been sponsored by the Government of The Netherlands through UNESCO-IHE, Delft, The Netherlands, and Nuffic/NPT-WREM Project, Butare, Rwanda.

CRC Press/Balkema is an imprint of the Taylor & Francis Group, an informa business

Published by:
CRC Press/Balkema
PO Box 447, 2300 AK Leiden, The Netherlands
e-mail: Pub.NL@taylorandfrancis.com
www.crcpress.com – www.taylorandfrancis.co.uk – www.balkema.nl

ISBN: 978-1-138-02657-5 (Taylor & Francis Group)

FOREWORD

The contribution to a better understanding on hydrological processes in a catchment for water resources planning and management in the region, that is characterized by very high competing demands (domestic vs. agricultural vs. industrial uses) was agreed since I started my PhD research in 2008. This was the time when Prof. Dr. Stefan Uhlenbrook was advertising the two PhD studies available to Rwandan young researchers, who hold Masters in a water related domain. Prof. Dr. Innocent Nhapi, former Manager of UR-WREM Project and Prof. Dr. Umaru G. Wali, the current Dean of the School of Engineering at UR, encouraged me to apply for this opportunity. I didn't hesitate and immediately applied because the subject was relevant and fitted with my hydrology and water resources background and interest. At that time, I was the Director at the Institute of Scientific and Technological Research (IRST) but I didn't regret the decision of leaving this position to continue my studies, as my dream was to become a researcher or a teacher with good qualifications. Academic career was my first choice and by chance the PhD in surface water hydrology was given to me after succeeding the interview, later combined with groundwater hydrology after my colleague Miss. Flora Umuhire left. I was asked to join the University of Rwanda (UR), Huye Caampus, which is former National University of Rwanda (NUR). I also accepted and was appointed in the Master's program of Water Resources and Environmental Management (WREM) as Assistant Lecturer and later promoted to Lecturer grade due to my publications/performance. My challenge was to first improve my knowledge in this domain to be able to teach in this program.

The WREM project aims to contribute to poverty alleviation and sustainable socio-economic development in Rwanda by stimulating solution-oriented research related to water resources and environmental management. I have been fully committed to contribute to the achievement of this mission. To contribute to this, my research focused on understanding the water resources of Rwanda and dominant hydrological process interactions in the specific meso-scale Migina catchment (257.4 km^2). The catchment has been equipped with hydrological instruments and after installation, rainfall, evaporation, runoff, hydrochemical and isotope data were observed over two years (May 2009 to June 2011). As a junior hydrologist, my challenge was to understand the hydrological processes in the targeted study area before the dominant hydrological process interactions could be assessed. However, due to my limited knowledge of hydrology and also lack of senior hydrologists in the whole country, my promoter Prof. Dr. Stefan Uhlenbrook decided to allocate to me two MSc research students (Harmen Van den Berg and Rutger Bolt from Vrije University of Amsterdam, The Netherlands) and we worked together intensively in this ungauged Migina catchment for four months from April 2009 onwards. In this short period, we have installed 13 rain gauges and three tipping buckets gauges, two evaporation pans, one weather station, five river gauging stations and eleven shallow piezometers for groundwater monitoring. During this time, a number of problems came such as some equipment like divers have stopped working and others were stolen. Beside this, some staff gauges were pushed away by unexpected large floods. Therefore, I had to regularly check and repair the equipments, replace divers with new ones and appoint guards on sites for equipments security. The data collected from this fieldwork and later, up to two years was used to understand the hydrological processes in a catchment. Due to short time series data, the recent collected data were checked for their quality/accuracy using long time series data recoded in and around the catchment before Rwandan genocide of 1994.

The Migina catchment was further developed to the water resources and environmental management research site of the National University of Rwanda. I hope and will do my best that the data collection will continue in this area to help the future BSc, MSc and PhD students carrying out their research in water management field and support decision makers for water resources planning and development.

During the last 5 years, I learned a lot about hydrological processes in a meso-scale catchment and found that cross-disciplinary knowledge and understanding constituted an important element of my individual and professional capabilities. I also gained huge personal experience by spending a lot of time in the field, where I got to know a large variety of people, and made many friends and learned about the importance of observing. I believe this experience will for sure contribute to the development of Rwanda as it has enriched my life and left me with many friends from different countries all over the world where I presented my findings. MSc students from the UR-WREM Programme will benefit from my experience as well as BSc students from Applied Sciences and Agriculture faculties in Rwanda.

Omar Munyaneza
UNESCO-IHE, Delft, The Netherlands
April 2014

ACKNOWLEDGEMENTS

This thesis is part of Water Resources and Environmental Management Project of University of Rwanda (WREM-UR) funded by the Netherlands Government through UNESCO-IHE and Nuffic/NPT PhD research fellowship. Additional funds for on-site implementation and data collection was facilitated by the Nile Basin Capacity Building Network Project (NBCBN) and the UR Research Commission facilitated to attend and present the research findings in different conferences. Moreover, I would like to say special thanks to UNESCO-IHE Institute for Water Education for their financial support and nice cooperation, Ms. Jolanda Boots (PhD Fellowship Officer at UNESCO-IHE) was remarkable and is sincerely acknowledged.

First of all, I would like to thank my promoter Prof. Dr. Stefan Uhlenbrook for all the constructive discussions and useful advice. Thank you for all the opportunities you gave me and making time in your always busy schedule to discuss my progress and give me helpful and quick feedback. Thanks you for your infinite effort to push and stimulate me to always focus on hydrological sciences. You never stopped to tell me "Omar, always think as a Hydrologist!". This thesis is the results.

I would also like to extend my sincere thanks to my supervisors/mentors from UNESCO-IHE, Dr. Shreedhar Maskey and Dr. Jochen Wenninger for all the help they provided me with and the guidance they kindly gave me during this study. Your significant contribution to this research is highly appreciated. I really admire your consistent encouragement to pursue excellence in every component of this study. I exceedingly benefited from your understanding of hydrological processes and modelling, critical thinking, instrumental set-up and technical writing ability. I am very very grateful for your dedication! Many thanks are due to my local PhD supervisor/mentor from UR, Prof. Dr. Eng. Umaru Garba Wali for his overall contribution in this research, especially for providing valuable professional insights, and guiding me during field work for more than 2 years. Your kindness and generous behavior will always be remembered.

I would like to express my thanks to the Rwandese family of Samuel Munyampeta and Jeanne Mukakalisa and your loved daughter Dolyne Munyampeta. You really let me feel at home in the Netherlands while away from my family. I really want to come back!

I would also like to thank the MSc and BSc students from University of Rwanda (UR), Huye Campus and Vriije University of Amsterdam, with whom I worked closely and whose research outputs contributed to this thesis: Anthony Twahirwa (MSc student), Felix Ufiteyezu, Gemma Maniraruta, Yves K. Nzeyimana, Jean Pierre Nkezabo (BSc students), and many others from UR; Harmen van den Berg and Rutger Bolt, MSc students from Vriije University of Amsterdam. Without you Harmen and Rutger, the instrumental set-up would have been too difficult for me. Dr. Jochen Wenninger also provided many contributions regarding the instrumental set-up and helping me in solving my diver equipment problems.

I am also thankful for all the support I received from my colleagues at UNESCO-IHE, especially the Water Science and Engineering Department and the laboratory (Fred Kruis, Ferdi Battes and Don van Galen) who helped me a lot with the many water samples that needed to be analysed. Lab technicians from UR laboratory such as Doris B. Gashugi, Mardoche Birori, Christine Niyotwambaza and Dieudonne, who helped me a lot as this was not my specialty, but I enjoyed the lab work with them. I recognize also the lab technician (Ms. Aloysia Kanzayire) from EWSA at Kadahokwa, Butare, for water samples analysis. Thanks!

In the Migina catchment, Rwanda, I would like to thank all the data collectors especially the teachers from the Primary Schools, where rain gauges were installed, who collected the valuable data, presented in this thesis and the guards who protected the delicate equipment. Special thanks to our driver Theobald Rudasingwa for taking us safely from place to place during the long days that we spent in the field for 2 years, and Prof. Dr. Eng. Umaru Garba Wali for assisting me in the field. Prof. Dr. Eng. Wali made me work in a good environment in my office at UR-Huye Campus and he always told me "work hard and don't wait for the deadline" (this thesis is the fruit of your advice!). Thanks to Prof. Dr. Eng. Sherif M. El-Sayed, former NBCBN Project Manager and Dr. Eng. Amel Azab, current Manager, you encouraged me to make some project proposals and get additional funds for this research. The conference Prize that I won for the best paper during the VII International Conference on Environmental Hydrology held at Cairo-Egypt on 27th September 2012, was the result of your encouraging financial support. This award gave me confidence that I was working on very interesting stuff, and a very nice recognition of my research. Dr. Jean Baptiste Nduwayezu, Director General of IRST (Institute of Scientific and Technological Research) and country coordinator of NBCBN, your collaboration will be always recognized. The contribution of Eng. Arsene Mukubwa during the training and advices provided to me regarding the use of HEC-HMS model is acknowledged. You helped me becoming a catchment modeler! Many thanks go also to Hon. Amb. Stanislas Kamanzi, Minister of Natural Resources in Rwanda (MINIRENA), for his encouragement. The two days visit in the Migina catchment accompanied with his technicians' team will always be recognized. You really made me feel proud of my research contribution for the development of Rwanda.

I would like to thank all the PhD students and staff at UNESCO-IHE for their friendship and nice discussions. Guy Beaujot and Mariëlle van Erven, Social Cultural Officers, for their cooperation, Jeltsje Kemerink, UR-WREM project manager and UNESCO-IHE Lecturer for your encouragement and negotiation made with nuffic for additional funds to my PhD budget. Thanks Dr. Marloes Mul for providing material and translating title and summary into Dutch and Dr. Ilyas Masih (my senior PhD fellow) for your advice and material provided and used in this thesis. Thanks to Mr. Abuba Selemani and my brother Said Sibomana for translating title and summary into Kinyarwanda, mother tongue. Thanks also to all PhD students at UNESCO-IHE from my home country: Mr. Abias Uwimana and Ms. Dominique Ingabire for your sacrifices to collect my data while in The Netherlands, Dr. Christian Birame Sekomo and Dr. Valentine Uwamariya for your coaching as my PhD fellow seniors.

Finally, I would also like to thank the person who is responsible for the start of my PhD study in the first place. Dr. Erik de Ruyter, Director of UR-WREM Project in the side of UNESCO-IHE, Prof. Dr. Innocent Nhapi, former Manager of UR-WREM Project, Prof. Dr. Elias Bizuru, Eng. Gonzalve Twagirayezu, Prof. Dr. U.G. Wali, Dean of the School of Engineering and my PhD local supervisor, Eng. Amri Birangwa, Mr. Said Bibomana, Mr. Yussuf Mugiraneza (brothers), Ms. Ziada Nyiraneza and Ms. Afisa Mugirwanake (sisters), Eng. Shabani Cishahayo and Eng. Issa Ndungutse (cousins), with your trust you convinced me to start a PhD and gave me the confidence to finish it. You gave me a family, love, and continuous support. I am most grateful for this.

Last but not least, I would like to thank my parents, my spouse's parents, wife Eng. Christine Uzayisenga, daughter Leïla Munezero and son Laïq Munyaneza for their deep love, good care and prayers for me, which provided me with the necessary support, comfort and energy to successfully complete this challenging venture. The sweet company of my wife Christine made this tough journey a very pleasant and memorable experience of my life to the extent that I feel very happy to dedicate this work to them all. You missed me for a long time and this thesis is the fruit of your patience and encouragement.

SUMMARY

Rapid population growth is increasing the water demand for domestic, agricultural and industrial uses and is causing water scarcity in Rwanda. There is also an increasing pressure on all natural resources including water. Identification of hydrological process interactions is essential for the proper management and assessment of the water resources within the catchments. It is also very important to understand the spatial and temporal distribution of water in a catchment at present and in the future. Lack of hydrological and climatic data in Rwanda, both in time and space, particularly after the genocide in 1994, is seriously impeding hydrological studies in this area.

The aim of this research is to explore the hydrological trends and climate linkages for catchments in Rwanda, with a particular focus on understanding dominant hydrological processes in the meso-scale Migina catchment, Southern Rwanda. Specifically, the study emphasizes the investigation of the relationship between trends in hydrologic variables and climate variability, quantifying the runoff components and identifying the dominant hydrological processes in a meso-scale catchment. The water resources availability in the meso-scale Migina catchment was also assessed in this research using catchment modelling.

A multi-methods including experimental and modelling activities was followed to achieve the research objectives. Different meteorological and hydrological instrumentations have been installed in the Migina catchment during the period April 2009 to July 2009. Meteorological data and river discharge measurements have been carried out and are still ongoing. In this research, a study on streamflow trends and climate linkages in the whole of Rwanda (26,338 km^2) was conducted for selected streamflow gauging stations over a long time period with minimum of 30 years (1961-2000).

Trends and the time of change points were investigated using the Mann-Kendall (MK) test and Pettitt test, respectively. The linkages between each of the climatic and hydrological variables were investigated using Pearson correlation. The results revealed significant trends for climatic and hydrological variables and an overall increasing trend of streamflow was detected for the longer period of data (1961-2000), however, a decreasing trend in the shorter period (1971-2000). Pettitt tests revealed that abrupt change points of most stations occurred in the 1980–1990s, which is related to the period of intensive human activities in Rwanda such as agriculture development (irrigation water use) and urbanization.

This thesis also attempts to quantify the runoff components and to identify the dominant hydrological processes in the Migina meso-scale catchment using hydrometric data and modern tracer methods. For the tracer method, we used two- and three-component hydrograph separation models (deuterium (^{2}H), oxygen-18 (^{18}O), chloride (Cl$^-$) and dissolved silica (SiO$_2$)). The results show that subsurface runoff is dominating the total discharge even during flood events. More than 80% of the discharge was generated by subsurface runoff for two investigated events (1st to 2nd May 2010 at the outlet of the Cyihene-Kansi sub-catchment and 29th April to 6th May 2011 at the outlet of the Migina catchment). This dominance of subsurface contributions is also in line with the observed low runoff coefficient values (16.7 and 44.5%) for both events. Hence, groundwater recharge mainly during the wet seasons leads to a perennial river system in the Migina catchment.

In the same Migina catchment, a simple rational method with area correction was used to predict river peak discharge of the Migina catchment. The agricultural land use dominates in the catchment (about 92.5%). The results show the weighted runoff coefficient of 0.25, time of concentration of 3.5 hours and the peak flow discharge (10 years return period) of 16 m^3 s^{-1}. Further, for assessing runoff and water resources availability on each sub-catchment level, we applied a semi-distributed conceptual hydrological model called Hydrologic Engineering Center-the Hydrologic Modelling System (HEC-HMS). The model is expected to assist as a tool for water resources planning and decision making processes in this catchment. The model was selected due to its capacity of analysing spatial variation of runoff generation characteristics, simplicity in setting-up, limited data requirements and free availability of the software.

The HEC-HMS version 3.5) was used with its soil moisture accounting, unit hydrograph, linear reservoir (for baseflow) and Muskingum-Cunge (river routing) methods. We used rainfall data from 12 stations and streamflow data from 5 stations, which were collected as part of this study over a period of two years (May 2009 and June 2011). The catchment was divided into five sub-catchments.

The model parameters were calibrated separately for each sub-catchment using the observed streamflow data. Calibration results obtained were found acceptable at four stations with a Nash–Sutcliffe Model Efficiency index of 0.65 on daily runoff at the catchment outlet. Due to the lack of sufficient and reliable data for longer periods, a model validation (e.g. split sample test) was not undertaken. However, we used results from tracer based hydrograph separation to compare our model results in terms of the runoff components. The model performed reasonably well in simulating the total flow volume, peak flow and timing as well as the portion of direct runoff and baseflow. We observed considerable disparities in the parameters (e.g. groundwater storage) and runoff components across the five sub-catchments, which provided insights into the different hydrological processes at sub-catchment scale. We conclude that such disparities justify the need to consider catchment subdivisions, if such parameters and components of the water cycle are to form the base for decision making in water resources planning in the catchment.

The knowledge generated here will be essential for decision-makers for setting national water policies and strategies for better water resources planning and management. The knowledge gained in this study will be transferred as much as possible to other Rwandan catchments to contribute to the 2020 vision of Rwanda and the Millennium Development Goals (MDGs). The Rwanda Vision 2020 is committed to reduce dependency on agriculture through employment diversification. This will hopefully reduce the pressure on water resources, given that agriculture accounts for more than 70% of the total water use at national level.

LIST OF ABBREVIATIONS AND ACRONYMS

DEM	Digital Elevation Map
DSS	Decision Support System or Data Storage System
EC	Electrical Conductivity [μS/cm]
EDPRS	Economic Development and Porverty Reduction Strategy
EGU	European Geophysical Union
EWSA	Energy, Water and Sanitation Authority
FAO	United Nations Food and Agriculture Organization
GDP	Gross Domestic Product
GIS	Geographic Information System
GPS	Global Positioning System
HBV	Hydrologiska Byråns Vattenbalansavdelning (Hydrological Bureau Water balance section)
HEC-GeoHMS	Hydrologic Engineering Center's Geospatial Hydrologic Modelling System
HEC-HMS	Hydrologic Engineering Center's Hydrologic Modelling System
IHP	International Hydrological Programme
IWRM	Integrated Water Resources Management
MC	Monte Carlo
MDG(s)	Millennium Development Goal(s)
MINAGRI	Ministry of Agriculture and Animal Resources (Rwanda)
MINALOC	Ministry of Local Government (Rwanda)
MININFRA	Ministry of Infrastructure (Rwanda)
MINIPLAN	Ministry of Planning (Rwanda)
MINIRENA	Ministry of Natural Resources (Rwanda)
MINITERE	Ministry of Land, Environment, Forest, Water and Mines (Rwanda)
NBCBN	Nile Basin Capacity Building Network
NELSAP	Nile Equatorial Lakes Subsidiary Action Program
NUR	National University of Rwanda
PUB	Predictions of Ungauged or poorly gauged Basins
REMA	Rwanda Environmental Management Authority
RMA	Rwanda Meteorological Agency
RMSE	Root Mean Square Error
RPSCM	Regional Project Steering Committee Meeting
RS	Remote Sensing
SCS	Soil Conservative Service
SENSE	Socio-Economic and Natural Sciences of the Environment
SMA	Soil Moisture Accounting
UN	United Nations
UNDP	United Nations Development Programme
UNEP	United Nations Environmental Programme
UNESCO	United Nations Educational, Scientific and Cultural Organization
UR	University of Rwanda
USDA-FAS	United States Department of Agriculture-Foreign Agricultural Service
USGS	United States Geological Survey
WMO	World Meteorological Organization
WRA	Water Resources Assessment
WREM	Water Resources and Environmental Management

TABLE OF CONTENTS

FOREWORD V

ACKNOWLEDGEMENTS VII

LIST OF ABBREVIATIONS AND ACRONYMS XI

TABLE OF CONTENTS XII

CHAPTER 1

INTRODUCTION 1

   1.1. BACKGROUND 1
   1.2. WATER RESOURCES IN RWANDA 2
   1.3. HYDROCLIMATIC DATA AVAILABILITY IN RWANDA 4
   1.4. DATA COLLECTION AND MANAGEMENT IN RWANDA 7
   1.5. DATA REPORTING AND SHARING SYSTEMS 7
   1.6. SPACE-TIME VARIATION OF HYDROLOGICAL PROCESS 8
   1.7. HYDROLOGICAL MODELLING 9
   1.8. PROBLEM STATEMENT AND OBJECTIVES 12
   1.9. THESIS OUTLINE 13

CHAPTER 2

METHODS AND MATERIALS 14

   2.1. INTRODUCTION 15
   2.2. STUDY AREA 15
      2.2.1. The Study Area Rwanda *15*
      2.2.2 The Migina catchment *17*
   2.3. INSTRUMENTATION 19
      2.3.1 Meteorology *19*
      2.3.2 River water *23*
      2.3.3 Groundwater *25*
      2.3.4 Hydrochemistry *27*
      2.3.5 Satellite imagery *28*

CHAPTER 3

STREAMFLOW TRENDS AND CLIMATE LINKAGES IN MESO-SCALE CATCHMENTS IN RWANDA 29

   3.1. INTRODUCTION 30
   3.2. DATA AND METHODS 30
      3.2.1. Selection of stations *30*
      3.2.2. Selection of hydrologic variables *32*
      3.2.3. Trends detection test *33*
      3.2.4. Change point detection *33*
   3.3. RESULTS 34
      3.3.1. Streamflow and climatic trends *34*
      3.3.2. Streamflow trends and climate linkages *37*
   3.4. DISCUSSION OF RESULTS 38
      3.4.1. Summary of trends *38*
      3.4.2. Relationship between hydrologic variables and climate variables *39*
   3.5. CONCLUSIONS 39

CHAPTER 4

IDENTIFICATION OF RUNOFF GENERATION PROCESSES USING HYDROMETRIC AND TRACER METHODS 41

   4.1 INTRODUCTION 42
   4.2 DATA AND METHODS 43

    4.2.1 Data collection *43*
    4.2.2 Field and laboratory methods *44*
    4.2.3 Hydrometric and tracer methods *44*
  4.3   RESULTS 46
    4.3.1 Rainfall-runoff observations for Itumba'10 & 11 seasons (March-May) *46*
    4.3.2 Results of hydrochemical tracer studies *48*
    4.3.3 Results of isotopes tracer studies *52*
  4.4 DISCUSSION 56
    4.4.1 Rainfall influence on runoff generation *56*
    4.4.2 Quantification of runoff components and processes in a meso-scale catchment ... *57*
  4.5 CONCLUSIONS 60

CHAPTER 5

**PREDICTION OF RIVER PEAK DISCHARGE IN AN AGRICULTURAL CATCHMENT IN
RWANDA** **61**

  5.1. INTRODUCTION 62
  5.2. DATA COLLECTION AND PROCESSING TECHNIQUES 63
    5.2.1. Physical characteristics of the catchment *63*
    5.2.2. Determination of Rainfall Intensity *63*
    5.2.3. Estimation of runoff coefficient *64*
    5.2.4. Determination of Peak Runoff *66*
  5.3. RESULTS AND DISCUSSIONS 67
    5.3.1. Data Processing and Results *67*
    5.3.2. Discussion *71*
  5.4. CONCLUDING REMARKS 72

CHAPTER 6

**ASSESSMENT OF SURFACE WATER RESOURCES AVAILABILITY USING CATCHMENT
MODELLING AND THE RESULTS OF TRACER STUDIES** **73**

  6.1. INTRODUCTION 74
  6.2. DATA AND METHODS 75
    6.2.1. Data *76*
    6.2.2. Methods (HEC-HMS 3.5 and HEC-GeoHMS 5.0) *77*
    6.2.3. Computation methods *77*
    6.2.4 Basin model setup and simulations *78*
    6.2.5 Calibration methods *79*
    6.2.6 Tracer techniques for model validating *80*
  6.3. RESULTS AND DISCUSSIONS 80
    6.3.1. Calibration Results *80*
    6.3.2 Simulated water budget components *83*
  6.4. CONCLUDING REMARKS 85

CHAPTER 7

CONCLUSIONS AND RECOMMENDATIONS **86**

  7.1. SUMMARY OF THE MAIN CONCLUSIONS 86
  7.2. RECOMMENDATIONS 88

SAMENVATTING **89**

INCAMAKE (SUMMARY IN KINYARWANDA) **91**

REFERENCES **94**

PUBLICATIONS BY THE AUTHOR **108**

ABOUT THE AUTHOR **110**

THE SENSE RESEARCH SCHOOL CERTIFICATE **111**

Chapter 1

INTRODUCTION

1.1. BACKGROUND

The world population has tripled in the twentieth century, while, over the same period, water use has increased about six fold. Currently, 1 billion people live in water-scarce or water-stressed regions, this number is expected to increase up to a factor 3.5 until 2025 (Wagener et al., 2008). The magnitude of this water scarcity and its variation in both space and time are largely unknown because of lack of hydro-climatological data (e.g. Oyebande, 2001; Kipkemboi, 2005). While the needs for hydrological and meteorological information are increasing, technical and human capacities are declining as noted by the reduction in number of hydrological and meteorological stations in Africa during the last 30 years (Bonifacio and Grimes, 1998).

Rwanda is also facing the problem of limited hydrological data with the rapid increase in population within its area of 26,338 km^2. The population density in the country is approximately 400 persons per km^2 (around 10,516,000 habitant; NISR, 2012), with an annual growth rate of about 3.5% (MINIPLAN, 2002). This makes Rwanda the most densely populated country on the African continent. With this population growth rate, the population will be doubled within twenty years. Rwanda needs to increase its food production levels to meet current and future demands and to reduce dependence on imports. In Rwanda water development for agriculture is a priority and especially in the southern part of Rwanda as reported by UNEP in 2005. This water development for agriculture cannot be met if the data and predictions are not available, and if good insights in water resources management are not well understood.

Sustainable management of water resources requires clear understanding of the catchment and its hydrologic processes. In order to understand this, the processes occurring in the catchment have to be understood at different scales, before the impact of these hydrologic processes can be assessed. Tracer methods provide suitable tools for investigating runoff generation processes (Uhlenbrook et al., 2002, 2008) and play a key role in catchment hydrology in identifying hydrological source areas and residence times (Soulsby et al., 2008). Hydrological processes within a catchment define how precipitation reaches the catchment outlet, how long water is stored in the river water, soil water and groundwater systems, as well as the hydrochemical composition of these components (Uhlenbrook et al., 2008, Wenninger et al., 2008). Understanding hydrological processes, in particular water flow pathways, source areas and residence times, is essential for predicting water quantities (including floods and low flows) and water quality in a catchment (Uhlenbrook et al., 2008).

Assessments of water quantity and quality in a catchment are needed for water managers and decision makers in view of Integrated Water Resources Management and Adaptive Management at catchment scale. However, such an assessment is still lacking in Rwanda to a large extent. As the population is mainly dependent on agriculture for their livelihood, Rwanda faces the challenge to provide all its inhabitants with enough food; today and tomorrow. Therefore, Rwanda is planning to reform the agriculture, for example by the introduction of different rice and maize crops and the construction of irrigation systems. The risk of severe water shortage during dry season as well as increasing pollution of the water resources is imminent. Knowledge about the water system, water availability and development of the water quality will be necessary to ensure human health nowadays and in the future (USEPA, 2012).

A number of recent studies in Rwandan catchments, particularly the Migina catchment, provide background information highly relevant to the issue of agriculture development, but lack information about dominant hydrological processes. Thus, a study on identification of dominant hydrological processes is needed to facilitate sustainable water resources management and planning in Rwanda. In this research we focused on the Migina catchment which is located in the southern part of Rwanda. The Migina catchment was selected because of three main reasons: i) to optimize the use of the catchment for agriculture (UR Research Agenda, 2007), because the economy of Rwanda is based on largely rain-fed agricultural (CIA World Factbook, 2012) and more than 90% of the population depend on agriculture for their livelihood (Davis *et al.*, 2010), ii) to implement the sustainable irrigation development plan as Migina is among identified priority catchments for irrigation purpose, given that agriculture accounts for more than 70% of the total water use at the national level, and iii) to set up an experiment catchment for hydrological study of the only existing public University in Rwanda which is located in the Migina catchment.

Due to the decline of hydro-meteorological stations countrywide caused by the civil war and genocide of 1994, the UNESCO-IHE Institute for Water Education started a capacity building programme in 2004, with the goal to train water specialists at the University of Rwanda (UR). One of the achievements of the programme is a Master of Science Degree Programme in Water Resources and Environmental Management (WREM) set up at UR and the author graduated in the first intake of 2007. This PhD research was a continuation of this capacity building project. Due to the above mentioned decline of hydro-meteorological stations, a hydrological fieldwork was carried out in the Migina catchment during the current study with the objective to equip the catchment with hydrometeorological measurement network. Several streamflow gauging stations were built, a meteorological tower was installed and rainfall stations were set up.

1.2. WATER RESOURCES IN RWANDA

A comprehensive study on the water resources in Rwanda was reported in FAO (2005), NBI (2005) and in PGNRE (2005a; 2005b). An overview of the state of the environment has been reported by REMA (2008, 2009b). RIWSP (2012b) summarizes all these reports and gave a clear picture of Rwanda water resources. Here, a brief overview of water resources of Rwanda will be presented, drawing largely on the above mentioned documents.

Generally Rwanda water resources have a direct influence on the quality of life of people, their health, livelihoods and their overall productivity (REMA, 2009). Thus, water is essential, not only to human life but for the environment, agriculture, industrial development, hydropower generation, transport, socio-economic development and poverty eradication. In Rwanda the occurrence of water resources is reflected by the existence of a network of wetlands in various parts of the country. Wetlands are generally represented by lakes, rivers and marshes associated with these lakes and rivers (MINITERE, 2005). Wetlands in Rwanda have been used in different ways and have a great role to play in the national economy growth. Main functions of wetlands in Rwanda include agriculture production, hydrological functions, biodiversity reservoirs, peat reserve, mitigation of climate change, leisure and tourism and cultural value (REMA, 2009).Rwanda has abundant rainfall and water resources, totaling $5 \times 10^9 \, m^3 \, a^{-1}$ (EarthTrends, 2003), with an average annual rainfall of about 1,200 mm a^{-1} (FAO, 2005). The temperature regime is more or less constant with a mean temperature of 16-17°C for the high altitude region, 18-21°C for the central plateau region and 20-24°C for the eastern plateau and lowlands (Twagiramungu, 2006). In 2000 the total water use was estimated at $150 \times 10^6 \, m^3/a$ of which the agricultural use accounts for 68%, the domestic sector 24%

and the industry 8% (FAO, 2005). Until early 1990s, the hydro-meteorological data network was dense and well distributed throughout the country (Fig. 1.1).

In Rwanda, water surfaces cover about 8% of the territory, or 2175 km^2 (NBI, 2005). A recent inventory of marshlands in Rwanda was conducted in 2008 and identified approximately 860 marshlands and flooded valley bottom land, covering a total surface of 2785 km^2, which corresponds to 9.4% of the country surface, 101 lakes covering 1495 km^2, of which more than 80% are part of Lake Kivu, and 861 rivers totaling 6462 km in length (REMA, 2008). Lake Kivu covers a surface area of around 1000 km^2 in the side of Rwanda (Munyaneza *et al.*, 2009a). In many regions of Rwanda, surface water is used as the only resource due to the fact that groundwater is not easily exploited. However, groundwater is the major source of clean water for rural areas used in the form of springs, which are estimated at around 22,300 (PGNRE, 2005b, Kente, 2011). These springs are the main source of drinking water in the Umutara region (eastern part of Rwanda) which is a semi-arid area with insufficient surface water and low potential of groundwater (PGNRE, 2005a). High potential zones are generally found in the alluvial valleys and the relatively thick soils in the south and the east, related to fissured granite and quartzite rock formations. These high potential groundwater zones are often related to terrain with modest or low slopes, where groundwater tables can be expected close to the surface (RIWSP, 2012b).

The summary of water resources in Rwanda is presented in Table 1.1 (Word Bank, 2006).

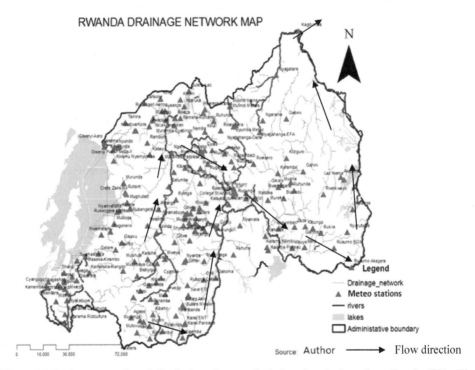

Figure 1.1 Drainage network and distribution of meteorological stations in Rwanda until early 1990s. The drainage network and distribution of hydro-meteorological stations after the 1994 Genocide is shown in Figures 1.2 and 1.3 (RIWSP, 2012a).

The water resources in Rwanda consist of numerous lakes and river systems. The Nile basin has six groups of lakes, namely: (i) the lakes of the north include lakes Bulera and Ruhondo and other small lakes; (ii) the lakes of the centre like Muhazi; (iii) the lakes of Bugesera: Rweru, Cyohoha (south and north), Kidogo, Gashanga, Rumira, Kilimbi, Gaharwa; (iv) the lakes of Gisaka: Mugesera, Birira and Sake; (v) the lakes of the Nasho basin: Mpanga, Cyambwe and Nasho; and (vi) the lakes of the Akagera National Park: Ihema, Kivumba, Hago, Mihindi, Rwanyakizinga (RIWSP, 2012a). In the Congo basin, Lake Kivu is the only lake. The major rivers include the Kagera, Akanyaru, Base, Kagitumba, Mukungwa, Muvumba, Nyabarongo, and Ruvubu in the Nile Basin and Koko, Rubyiro, Ruhwa, Rusizi, Sebeya in the Congo Basin (NBI, 2005; Kabalisa, 2006; REMA, 2009).

Table 1.1 Water Resources in Rwanda (Source: Word Bank, 2006)

Category		Units	Quantity
Internal Renewable Water Resources (IRWR), 1977-2001	Surface water (SW) produced internally	km³ a⁻¹	5
	Groundwater (GW) recharge	km³ a⁻¹	4
	Overlap (O) shared by groundwater and surface water	km³ a⁻¹	4
	Total per capita IRWR, 2001	m³ cap⁻¹ a⁻¹	638
Natural Renewable Water Resources (includes flows from other countries)	Total, 1977-2001	km³ a⁻¹	5
	Total per capita IRWR, 2002	m³ cap⁻¹ a⁻¹	638
	Annual river flows:		
	- From other countries	km³ a⁻¹	X
	- To other countries	km³ a⁻¹	X
Water withdrawals (Year 1993)	Total withdrawals	km³ a⁻¹	0.8
	Withdrawals per capita	m³ cap⁻¹ a⁻¹	141
	Withdrawals as a percentage of actual renewable water resources	%	22
	Withdrawals by sector (as a percent of total)*:		
Water withdrawals (Year 2000)	- Agriculture	%	68
	- Industry	%	8
	- Domestic	%	24

Totals may exceed 100 percent due to groundwater drawdowns, withdrawals from river inflows

Access to water supply in 2008 was at national level 71% with urban coverage of 76% and rural drinking water coverage of 68% (Kente, 2011). The water sanitation for the whole country was approximately 85% with only 58% supply which was meeting suitable hygienic standards of WHO. The study of Kente (2011), on overview of water resources in Rwanda, found that:

a) Water resources in Rwanda can easily satisfy the needs of the population, if used efficiently. However, their protection is also necessary;
b) Policies and laws governing the water sector have one goal of sustainable environment, but they need to be implemented effectively;
c) Different stakeholders are involved in the water sector include government institutions (e.g. MINIRENA, REMA, MININFRA, EWSA, RNRA, MINAGRI) professional groups, facilitators and donors and local communities;
d) At national level, there are important programmes, Vision 2020, EDPRS, which give major orientations for social and economic development and management of natural resources by the horizon 2020;
e) The level of collaboration between the water and other linked sectors is still weak; and
f) Data and information sharing is unsatisfactory which can lead to unnecessary repetitive interventions.

1.3. HYDROCLIMATIC DATA AVAILABILITY IN RWANDA

Meteorological data in Rwanda have been collected since the colonial times by Germans in 1907 and later Belgians in 1917, when the country was part of the territory of the great Belgian Congo (Dushimire, 2007). The first meteorological station was established in the year 1907 at Save Station,

5

which is located in the Migina catchment in Southern part of Rwanda (Fig. 2.4). At this station, rainfall data is available from 1910 up to 1992, although with many gaps. The Migina catchment is the main focus area of this research where a new hydro meteorological network was installed in May 2009 for the purpose of this research. A small number of meteorological stations were established in the 1930s, at different location (see Fig. 1.1). In 1955, the Office of Meteorology was established taking control of all meteorological stations. The total number of meteorological stations managed by this office once reached 195, but was reduced to 147 stations in 1990 (MINITERE, 2005) and 136 stations in 1993 (Dushimire, 2007). Many of these stations were destroyed during the period of civil war and genocide between 1990 and 1994; the majority of them in 1994 (Munyaneza *et al.*, 2010), and some of them have only recently been rehabilitated.

For this study, historical climatic data were collected from either the data base developed by SHER Company in 2004 or the Rwanda Meteorological Office. Up to 2008, only eleven of the above mentioned meteorological stations were working around the country, and the available data are not continuous and complete. In 2013, Rwanda Meteorological Agency (RMA) had a network of observations stations around the country that monitor the climate and weather composed of 8 synoptic stations, 5 agro-synoptic stations, 39 automatic weather observing stations and 71 operational rainfall observing stations (see Fig. 1.2 and MINIRENA, 2013). The variables measured at climatological stations include temperature, rainfall, wind speed, wind direction, atmospheric pressure, soil moisture, and cloud cover (RMA, 2013). New data are also needed to be collected in this study for more accurate water balance studies, assessment of water resources availability in the Migina catchment.

The selection of climatic stations for this study was based on the length of the available data series and their spatial distribution over the country (see Chap. 3).

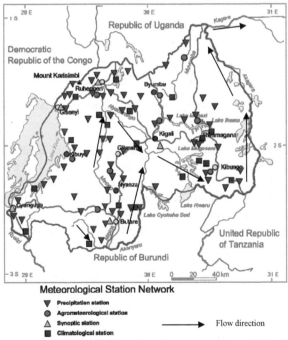

Figure 1.2 Meteorological stations rehabilitated or installed after the genocide of 1994 till 2013 around the country (RIWSP, 2012a).

In the data base developed by SHER in 2004, 35 river gauging stations were identified around the country with daily water level and corresponding discharge. Data are available from 1961 up to 2000 (between 5 to 40 years of data are available). Similar to meteorological stations, many of these hydrological stations were destroyed during the period of civil war and genocide between 1990 and 1994 as well, and some have only recently been rehabilitated.

In this research, new data from 5 gauging stations installed in the Migina catchment in April 2009, were used (after cross checking of their quality) for water balance and rainfall-runoff modelling (Chapter 6).

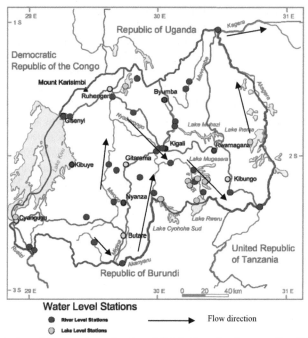

Water Level Stations

● River Level Stations   ⟶ Flow direction
◐ Lake Level Stations

Figure 1.3 Rwanda hydrometric observing network and distribution after the genocide of 1994 to 2013 around the country (RIWSP, 2012a).

There are also data available from four additional gauging stations installed in the Migina catchment by AGRAR-UND HYDROTECHNIK Company in 1991 (AGRAR, 1993), whereby only one year of data is available in hard copies from these AGRAR gauges. All those stations located in the Migina catchment were not working at the beginning of this research. However, for conducting better hydrological analysis in this catchment, re-installation or installation of new stations and collection of new data was needed. Those four river gauging were rehabilitated and made operational since May 2009 for this research. Nowadays, 5 river gauging stations are functioning in the Migina catchment (see Fig. 2.4).

When we started this research in 2008, only 22 gauging stations were reinstalled by the Ministry of Natural Resources (MINIRENA) through its National Water Resources Management Project, and were functioning in the Rwandan catchments (NWRMP, 2008; Munyaneza *et al.*, 2010). None are located in the Migina catchment.

In 2013, the hydrological observation network is composed of 41 hydrometric stations (Fig. 1.3); of which 31 are located at river sites and the remaining 10 measure lake water levels. The department of IWRM in Rwanda Natural Resources Authority (RNRA) is currently in the process of upgrading the observing network infrastructure. 12 new automatic water level stations were constructed/rehabilitated by January 2014 but not yet operational. This will provide water level observations at one-hour frequency. Presently, only 17 river gauging stations with rating curves are available. But these curves are derived from discharge measurement dating back to before 1990 (RNRA, 2012). As a result, the discharge ratings may not be valid anymore to provide reliable discharges for present observations. Moreover, the rating curves need to be updated regularly because major events such as floods may cause sudden changes in the cross section and the river bed conditions (Kennedy, 1984, De Laat and Savenije, 2002). Recently RNRA has equipped 10 key stations with OTT Thalimedes automatic water level measurement systems, which measure and store water levels at hourly interval. Data from automatic stations are downloaded continuously after some time for utilization and river discharge measurements are carried out for available rating curves checking and updating (RNRA, 2012).

Manual flow discharge measurements are carried out by RNRA using current meter. In addition, recently ADCPs (Acoustic Doppler Current Profiler) are used because they are very useful for large rivers. However, more stations are needed to be installed to improve the accuracy of so far data collected.

1.4. DATA COLLECTION AND MANAGEMENT IN RWANDA

Nowadays, readings from meteorological instruments are regularly taken and transmitted to Rwanda Meteorological Agency (RMA) headquarters where quality control is performed. Data are received from the field hydrological observers in a web system or downloaded quarterly from loggers. Data are checked for quality control by comparing levels from data loggers with manual staff gauge readings and processed before entry in database. So far, RNRA uses a simple Aqualium database system for the storage of its observation data. After quality control is performed, the data are sent to the Global Telecommunications System in Nairobi, analyzed for forecasts and stored in Meteo Rwanda's data archive. There are comprehensive meteorological data records from 1960s, when Meteo Rwanda was established, until around 1994.

1.5. DATA REPORTING AND SHARING SYSTEMS

There are bulletins produced by Rwanda Meteorological Agency (RMA) and shared with stakeholders and researchers. Agro-meteorological bulletins provide the following information: weather summary, maps of rainfall and percent of normal rainfall, weather forecast, vegetation conditions and weather impact on agriculture, and expected weather impacts on agriculture. Bulletins are produced every 10 days and the ones on the third decade of each month are prepared together with the monthly or seasonal bulletins. Those bulletins are presented in form of hyetograph, which are followed by some interpretations. Those hyetographs seem reliable but observers always need to be trained on the way of proper data collection. Good news is that the country is evolving towards automatizing stations and recently around 15 were installed in cooperation between Rwanda Environment Management Authority (REMA) and Rwanda Meteorological Agency (RMA) and data will be recorded and sent directly to the servers. Therefore, the data should be more reliable than it was before.

There are also weather stations installed with automatic rain gauges (tipping buckets) such as the one installed at the Center of GIS located at Butare, southern Rwanda, where this study is located, and the other one installed in Nyungwe forest in southern Rwanda. These two meteo stations have been installed in 2006 and up to date data are available (rainfall intensity and other climate data).

Hydrological bulletins are produced by Rwanda Natural Resources Authority (RNRA) in its department of Integrated Water Resources Management (IWRM) on a monthly basis. The bulletins are shared with stakeholders and published to the RNRA website for easy access by users (RNRA, 2012). The production of this monthly bulletin has stopped in February 2013 and the RNRA is now producing it every 3 months. Hydrological data are shared to users free of charge upon filling the form available on the RNRA website.

1.6. SPACE-TIME VARIATION OF HYDROLOGICAL PROCESS

Spatial variability of rainfall is often considered as a major source of temporal variability in the resulting basin hydrograph. Since direct measurements are not available, this must be verified through a modelling approach, provided adequate data are available (Obled *et al.*, 1994). Hence, the choice of a model is determined by the purpose of the model and the availability of data (see Chap. 6). The spatial scale is very much linked to the temporal scale through the residence times of water in the catchments (Jothiyangkoon *et al.*, 2001).

The rainfall in Rwanda generally occurs throughout the year, with quite some spatial and temporal variability. Eastern and southeastern regions are more affected by prolonged droughts while the northern and western regions experience abundant rainfall that at times cause erosion, floods and landslides. The spatial variability has been attributed to the complex topography and the existence of large water bodies within the Great Lakes Region.

Observations and analysis from existing data shows that over the last 30 years, some parts of Rwanda have experienced unusual irregularities in climate patterns including variability in rainfall frequencies and intensity, persistence of extremes like heavy rainfall in the northern parts and drought in the eastern and southern parts (REMA, 2009b).

Analysis of rainfall trends in Rwanda shows that rainy seasons are tending to become shorter with higher intensity (REMA, 2009b). This tendency has led to decreases in agricultural production and events such as droughts in dry areas; and floods or landslides in areas experiencing heavy rains. Heavy rains have been being observed especially in the northern and the western province. These heavy rains coupled with a loss of ecosystems services resulting from deforestation and poor agricultural practices have resulted in soil erosion, rock falls, landslides and floods which destroy crops, houses and other infrastructure (roads, bridges and schools) as well as loss of human and animal lives. On the other hand the eastern region of the country has been experiencing rainfall deficits over the last decades (REMA, 2009b). Observations between 1961 and 2005 showed that the period between 1991 and 2000 has been the driest since 1961. These observations showed a marked deficit in 1992, 1993, 1996, 1999 and 2000 with rainfall excesses in 1998 and 2001 (MINITERE, 2006).

The improved understanding of the hydrological processes will be beneficial to quantify the runoff components and to identify the dominant hydrological processes in a meso-scale catchment (e.g. Uhlenbrook *et al.*, 2002, 2007 and 2008). Additionally, the improved understanding of the hydrological processes will improve the hydrological modelling and provide the basis for the hydrological conceptualization, which is required to understand the implications on the modelling innovations.

Recently, Renno *et al.* (2008) showed that the Height Above the Nearest Drainage (HAND) is a much more powerful tool to distinguish hydrological landscapes than mere elevation. Savenije (2010) defined wetland, hillslope and plateau in hydrological terms as follows: "... *a wetland stands for a hydrological landscape element where saturation excess overland flow (SOF) is the dominant runoff mechanism. Likewise the term hillslope stands for a hydrological landscape element where storage excess subsurface flow (SSF) is the dominant runoff mechanism. Plateau stands for hydrological landscape elements with modest slope where the groundwater table is deep and where the dominant mechanism is evaporation excess deep percolation (DP)*". We tried to identify the Migina hydrological processes based on the above definitions (Chapter 4).

Unfortunately, when I started this reseach in 2008, Migina catchment was ungauged as many catchments of Rwanda and even the gauged catchments have unreliable data sets. Data challenge is not only in Rwanda, but also many other countries worldwide are ungauged or poorly gauged. In addition, existing measurement networks are declining worldwide (Sivapalan *et al.*, 2003) and this can lead to uncertainty. Because to this observed challenge, a new initiative of PUB was launched by the International Association of Hydrological Sciences (IAHS), aimed at formulating and implementing appropriate science programmes to engage the scientific community, and improve their capacity for making predictions in ungauged basins (Sivapalan *et al.*, 2003). It should be noted that the main focus of the PUB initiative was on predictions in ungauged basins, and also data to support these predictions (Hrachowitz *et al.*, 2013b; Savenije and Sivapalan, 2013). In response to the limitations of applicable approaches, flexible models (e.g. FLEX model) have received support during the PUB Decade (e.g. Beven, 2000; McDonnell 2003; Savenije 2009; Hrachowitz *et al.*, 2013b). These flexible models can allow consistent comparison and testing of alternative model hypotheses.

1.7. HYDROLOGICAL MODELLING

For many years, modelling tools have been available to simulate spatially distributed hydrological processes. These tools have been used for testing hypotheses about the behaviour of natural systems, for practical applications such as erosion and transport modelling, and for simulation of the effect of land use or climate change. However, so far the quality of the simulations and spatial process representations has been difficult to assess because of a lack of appropriate field data (Grayson and Blöschl, 2000).

In recent years, there have been several major field experiments in research catchments, aimed specifically at improving our understanding and modelling capability of spatial processes. This research used some of those studies, and field work has been conducted in Rwanda for better understanding of spatial hydrological processes (see Chap. 2).

Hydrologic models are increasingly used to support decisions at various levels and guide water resources policy formulation, management and regulations (Magoma, 2009). Hydrological models are often successful in simulating basin discharge, even if very simplified or sometimes unrealistic concepts are used (Uhlenbrook *et al.*, 1999). This is caused by the fact that these models are often over parameterized (Beven, 1989; Jakeman and Hornberger, 1993) and that they are validated by only a single measure: the simulated runoff (Hoeg *et al.*, 2000). Hoeg *et al.* (2000) stated that: "*An investigation on the dominating runoff generation processes in the catchment before a model is set up can reduce such uncertainties. On the one hand, knowledge of processes and flow pathways is crucial for evaluating the vulnerability of surface and groundwater systems (Leibundgut, 1998). On the other hand, knowledge of runoff generation processes helps to develop and validate the concepts of hydrological models*". Hydrochemical and isotope tracers were used in this study and were found to be suitable tools for investigating runoff generation processes (e.g. Chapter 4 of this thesis).

Many attempts have been made to classify rainfall-runoff models (e.g., Clarke, 1973; Todini, 1988; Chow *et al.*, 1988; Singh, 1995; Refsgaard, 1996; Wagener *et al.*, 2007; Wagener *et al.*, 2008; Shrestha, 2009; Solomatine, 2011). The classifications are generally based on the following criteria (Shrestha, 2009): "*(i) the extent of physical principles that are applied in the model structure; and (ii) the treatment of the model inputs and parameters as a function of space and time*". According to the first criterion (i.e. physical process description), a rainfall-runoff model can be classified into: (i) physically based models, (ii) conceptual models, and (iii) data-driven models.

The physically based models are based on the general principles of physical processes (e.g. continuity, momentum and/or energy conservation) and describe the system behaviour in as much detail as conceivable. The state and evolution of the system is described using state variables that are functions of both space and time. These variables allow physical meaning and most of them are measurable, though often not at catchment scale (Abbott *et al.*, 1986). The principles used in such models are assumed to be valid for a wide range of situations including those that have not yet been observed (Guinot and Gourbesville, 2003; Graham and Butts, 2005). SHE/MIKE-SHE is a typical example of a physically based hydrological modelling system (Abbott *et al.*, 1986a, b).

MIKE-SHE (Systeme Hydrologique Européen) can treat many water management issues in an integrated fashion, at a wide range of spatial and temporal scales. It has been used for the analysis, planning and management of a wide range of water resources and environmental and ecological problems related to surface water and groundwater. However, there are important limitations to the applicability of such physics based models. For example (Graham and Butts, 2005): i) such models require a significant amount of data; ii) the relative complexity of the physics-based solution requires substantial execution time and may lead to over-parameterized descriptions for simple applications; and iii) a physics-based model attempts to represent flow processes at the grid scale with mathematical descriptions that are valid for small-scale experimental conditions. Thus, a complete, physics-based flow description for all processes in one model is rarely possible. Graham and Butts (2005) suggested that a sensible way forward is to use physics-based flow descriptions for only the processes that are important, and simpler, faster, less data demanding methods for the less important processes.

The data-driven (also called empirical) models, involve mathematical equations that have been not derived from the physical processes in the catchments but from an analysis of the concurrent input and output time series (Solomatine, 2011). Typically such models are valid only within the boundaries of the domain where data is given (Price, 2006). Artificial Neural Networks is an example of data-driven model. Czop *et al.* (2011) tested the formulation and identification of First-Principle Data-Driven models. They found that in case the multiple parameters are simultaneously estimated, adjusting a model to data is in most of the time a non-convex optimization problem, and the criterion function may have several local minima.

Conceptual models are generally composed of a number of interconnected storages, which are recharged through fluxes of rainfall, infiltration or percolation and depleted through evaporation and runoff assembling the real physical process in the catchment (Shrestha, 2009). The equations used to describe the processes are semi-empirical, but still with a physical basis. The model parameters cannot usually be assessed from field data alone, but have to be obtained through calibration.

The conceptual models are by far the most widely used models for most practical applications. Comparison results of 10 different conceptual models used in the 1960s for operational hydrological forecasting are presented in WMO (1975). More comprehensive descriptions of a large number of conceptual models are provided in Singh (1995). However, there are many conceptual models with different levels of physical representational and varying degree of complexity. Crawford and Linsley (1966) are credited for the development of the first major conceptual model by introducing the well-known Stanford Watershed Model. Numerous other widely used conceptual models include Sacramento Soil Moisture Accounting model (Burnash *et al.*, 1973), NAM model (Nielsen and Hansen, 1973), TOPMODEL (Beven and Kirkby, 1979), TANK model (Sugawara, 1967, 1995), HBV model (Bergström and Forsman, 1973), FLEX model (Savenije, 3013; Gharari *et al.*, 2013; Gao *et al.*, 2013). A brief description of several conceptual models is given in an early work by Fleming (1975), and a description of flexible models is recently given by Fenicia *et al.* (2008a,b).

Gao *et al.* (2013) applied the flexible model (FLEX-Topo) in nested catchments in the Upper Heihe basin in China by comparing three model structures: a lumped model (FLEX^L), a semi-distributed model (FLEX^D), and a conceptual model (FLEX^T). They found that the conceptual model, FLEX^T, performs better than the other models in the nested sub-catchment validation and it is better transferable due to its flexibility in model structure. Gharari *et al.* (2013) also applied FLEX-Topo model with the aim to efficiently exploiting the complexity of a semi-distributed model formulation. They found that more complexity of models allows more imposed constraints. However, they found that a constrained but uncalibrated semi-distributed model can predict runoff with similar performance than a calibrated lumped model. They concluded that, if not warranted by data, models with higher complexity suffer from higher predictive uncertainty and may include many processes. Gharari *et al.* (2013) noted that to make more efficient the use of model sensitivities to these constraints, FLEX-Topo framework needs to be evaluated in the future with additional internal information, such as groundwater dynamics (e.g. Seibert, 2003; Fenicia *et al.*, 2008a) or tracer data (e.g. Birkel *et al.*, 2011; Capell *et al.*, 2012; Hrachowitz *et al.*, 2013a). Moreover, the suitability of model structures and parameterizations is assigned to the different hydrological response units, *HRUs*, (Fenicia *et al.*, 2011; Gharari *et al.*, 2013).

Conceptual models have been also applied in the African catchments as well as in Rwanda. For example: Githui *et al.* (2009) used SWAT model to simulate stream flow in Western Kenya. Results revealed important rainfall-runoff linear relationships for certain months that could be extrapolated to estimate amounts of stream flow under various scenarios of change in rainfall. They recommended that, if all other variables like land cover and population growth, were held constant, a significant increase in stream flow would be expected in the coming decades due to the consequence of increased rainfall amounts. Sang (2005) also applied SWAT model in Nyando Basin in Kenya and observed that an increase of rainfall by 15% would increase peak flow from 111 m^3 s^{-1} to 159 m^3 s^{-1}. Magoma (2009) examined the applicability of SWAT in the Rugezi wetland catchment in Rwanda (197 km^2). He found that the simulated flows at Rusumo gauging station comply with the measured flows.

In this study (Chap. 6), the HEC-HMS (Hydrologic Engineering Center - Hydrologic Modelling System, version 3.5) is used, which is a conceptual semi-distributed hydrological model. The HEC-HMS was designed to simulate the rainfall-runoff processes for the catchment systems (USACE, 2008, Scharffenberg and Fleming, 2010). Its design allows applicability in a wide range of geographic areas for solving diverse problems including large river basin water supply and flood hydrology, and small urban or natural catchment runoff (Merwade, 2007). The HEC-HMS model was set up in the meso-scale Migina catchment (257.4 km^2) located in southern Rwanda to simulate the catchment discharge and to assess spatio-temporal availability of water resources (see Chap. 6). Simplicity in setting-up, low data demand for running simulations and the fact that it is public domain software are some of the reasons for choosing this model. Computations in HEC-HMS include loss calculations, conversion of extreme rainfall to runoff, baseflow estimation, routing in reaches and reservoirs (Sardoii *et al.*, 2012).

HEC-HMS has been successfully applied in many catchments worldwide. For example: Christopher and Yung (2001) used HEC-HMS to perform a grid-based hydrologic analysis of a catchment. They compared distributed, semi-distributed and lumped models and reasonable contribution of flood observation and runoff volume. Fleming and Neary (2004) used successfully HEC-HMS as a tool for continuous hydrologic simulation in the Cumberland River basin. Neary *et al.* (2004) applied the HEC-HMS model for continuous simulation by comparing streamflow simulations using basin-average gauge and basin average radar estimates. Cunderlik and Simonovic (2005) also used the continuous simulation version of the HEC-HMS model to describe the main hydroclimatic processes in the Ontario River basin. Chu and Steinman (2009) carried out continuous hydrologic simulations by applying HEC-HMS to the Mona Lake watershed in west Michigan. Bouabid and Elalaoui (2010) used HEC-HMS for assessing the impact of climate change on water resources in the Sebou Basin in Morocco. Boyogueno *et al.* (2012) applied HEC-HMS for the prediction of Flow-Rate in Sanaga Basin in Cameroon.

When we started this research in 2008, we did not find any research which has been conducted in Rwandan catchments using HEC-HMS model. However, this study used HEC-HMS version 3.5 for testing its applicability in a meso-scale catchment and to inform water resources planning and decision making for better use of Migina catchment.

1.8. PROBLEM STATEMENT AND OBJECTIVES

Knowledge and understanding of different hydrological processes and their interactions with climatic variables are essential for the present and future assessment of water resources availability. These are also pre-requisites for improved planning and sustainable management of water resources (WaterNet, 2008; Masih *et al.*, 2011). Unfortunately, there are some critical issues that many African catchments are facing, which include poor water resources management and planning, climate variability and change, water scarcity because of rapid riparian population growth and urbanization, and lack of adequate hydro-climatic data.

In Rwanda, the main problems include: (i) lack of sufficient studies in this area; (ii) lack of sufficient data particularly in the post 1994 period because of destruction of hydro-meteorological stations together with many missing historical data sets; and (iii) lack of human resources with skills in hydrology and water resources management.

Climatic and hydrological data are key components of water resources management and assessment because sustainable water resources planning and management require data to enable quantification of water quality and quantity (Oyebande, 2001; Zhang *et al.*, 2011). Lack of adequate hydro-climatological data causes uncertainty in the design, management and assessment of water resources systems. The review of hydro-climatic data availability and analysis of hydro-climatic variability in Rwanda is required for identification of hydrological processes and assessment of national water resources availability.

Some recent studies in Rwandan catchments, particularly in the Migina catchment provide background information highly relevant to the issue of agricultural development, but in those studies, the information about dominant hydrological processes is missing. Lack of such information has negative impacts on the distribution of water resources in time and space for various uses including agriculture (WRPM, 2006). Thus, new techniques such as tracer tests and modelling need to be developed to gain a better understanding of water resources assessment and management (Shadeed, 2008).

As a solution for food security and poverty alleviation, Rwandan marshlands are being converted for intensive agricultural activities (World Bank, 2008). However, these goals cannot be met, if there are insufficient data and unknown water resources availability in the catchments. That is why rehabilitation of the gauging stations, the review of hydro-climatic data availability and trend analysis of existing hydro-climatic data are a priority. Furthermore, the collection of new data to enable feasibility studies to be carried out, for potential water resources assessment, are utmost important.

The main objective of this thesis is to explore the hydrological trends and climate linkages for the catchments in Rwanda, with a particular focus on understanding dominant hydrological processes in the meso-scale Migina catchment.
The specific objectives are to:
  i.   Investigate the relationship between trends in hydrologic variables and climate variability and climate change in meso-scale catchments in Rwanda;
  ii. Quantify the runoff components and identify the dominant hydrological processes in the Migina meso-scale catchment using hydrometric and innovative tracer methods; and
  iii. Assess the spatio-temporal availability of water resources in a meso-scale catchment using catchment modelling.

1.9. THESIS OUTLINE

In Chapter 1 the importance of the problem selected is established and a clear statement of the problem is given. Literature review on hydrological process and hydrological modelling are synthesized in this part. The current situation of existing hydro-meteorological stations in Rwanda is introduced in this chapter. Additionally, the problem statement and objectives of this work are given in this chapter.

In Chapter 2 the materials and methods which are used in the study are presented. A description of the study area is presented in this chapter. The detailed hydrological and meteorological instrumentation network that was setup in the study site Migina is described. The description of data characteristics, procedures used for achieving the research objectives and assumptions made thereof are also explained in this chapter.

In Chapter 3 the relationship between trends in streamflows and hydrologic variables are investigated (Masih *et al.*, 2010). Rainfall and temperature data are used to detect trends in climatic variables and their correlations with streamflows. Trends and change points were investigated and summarized in this chapter.

The runoff generation processes in a meso-scale catchment are indentified in Chapter 4. The use of hydrochemical and isotope parameters for separating streamflow into different runoff components is examined.

In Chapter 5 a simple rational method is developed to assess the high peak flow discharge in an agricultural catchment. A land use map and hydrological soil group map of a meso-scale catchment are digitized to estimate the runoff coefficient of catchments to calculate a weighted runoff coefficient. The rainfall intensity of Migina catchment is determined and the time of concentration is computed in this chapter.

In Chapter 6 water resources availability in a meso-scale catchment are assessed using catchment modelling. The spatial and temporal distribution of water to improve the management of water resources is presented. Hence, the knowledge of water resources availability in a catchment is provided in this chapter. The application of the HEC-HMS model in a meso-scale catchment is detailed in this chapter.

The thesis ends with the conclusions in Chapter 7, where all findings are summarized and some recommendations are given for future research.

Chapter 2

METHODS AND MATERIALS

....

The proper implementation of Integrated Water Resources Management (IWRM) plans requires the collection and analysis of hydrological and meteorological time series. This paper introduces the current situation of existing hydro-meteorological stations in Rwanda and describes a detailed hydrological and meteorological instrumentation network that was setup in the Migina catchment (257.4 km². This includes 13 rain gauges and three tipping buckets gauges, two evaporation pans, one weather station, five river gauging stations and eleven shallow piezometers for groundwater monitoring. Based on the data collected, rating curves for the 5 river gauging stations have been established. During the period from May to December 2009, the maximum rainfall of 52.5 mm d⁻¹ was observed in November at Mpare rainfall station (1691 m a.s.l.). The highest peak flow was observed on 19 November 2009 at the outlet of the Migina catchment at Migina River (4.8 m³ s⁻¹), and the lowest flow was observed on 8 August 2009 at Munyazi-Rwabuye River (0.0002 m³ s⁻¹), which is located at upstream of the Migina catchment. In future, this catchment will be developed further to the water resources and environmental management research site of the National University of Rwanda.

....

Based on: Munyaneza, O., Uhlenbrook S., Wenninger, J., van den Berg, H., Bolt H. Rutger, Wali G.U. and Maskey S., 2010.
Setup of a Hydrological Instrumentation Network in a Meso-Scale Catchment- the case of the Migina Catchment, Southern Rwanda. Nile Water Sci. Eng. J., 3(1): 61-70.

2.1. INTRODUCTION

The analysis of climatic and hydrological trends was carried out for all Rwandan catchments. The detailed analysis of hydrological process interactions and the assessment of water resources availability were focused on the meso-scale Migina catchment.

During the study period, a hydrological measurement network was built up in the meso-scale Migina catchment. In total five gauging stations, one meteorological station, eleven piezometers, twelve totalizers and four tipping buckets were installed throughout Migina catchment. In addition there was already a meteorological station present at the CGIS-centre in Butare (located at upstream of the catchment).

The study area Rwanda and the Migina catchment case study area are discussed below.

2.2. STUDY AREA

2.2.1. The Study Area Rwanda

Rwanda is characterized by a tropical temperate climate due to its high altitude, with a surface area of 26,338 km^2 of which 2,175 km^2 is covered by water (NBI, 2005). It is dominated by two wet and two dry seasons (Verdoodt and Van Ranst, 2003). Altitude ranges from 950 m a.s.l. at the Rusizi River in the south-west to 4,507 m a.s.l. at Mount Karisimbi in the north-west (RIWSP, 2012b). The country is divided into two main basins (Fig. 2.1): Congo River and Nile River. The Congo River basin located to the west and occupies 20% of the country and drains 10% of the country's waters towards Lake Kivu. In the south-west, the Rusizi River draining Lake Kivu towards Lake Tanganyika forms the south-west border with the Democratic Republic of the Congo. The Nile River basin located to the east occupies 80% of the Rwanda and drains 90% of the nation's water resources towards Lake Victoria via the Kagera River (Munyaneza *et al.*, 2011a), the major river in the south and eastern part of Rwanda (FAO, 1997; MINITERE, 2005). The Nile River basin in Rwanda covers an area of 19,876 km^2, which is 0.6% of the whole Nile River basin (FAO, 1997). The basin has an elevation mostly around 1,200-1,600 m a.s.l., which may rise up to 2,500 m a.s.l. in the west and reaches to 4,500 m a.s.l. in the north (Mutabazi *et al.*, 2004).

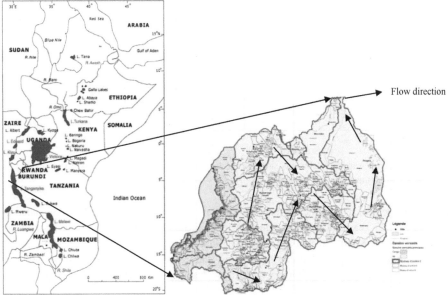

Figure 2.1 The location of Rwanda in the Nile basin Countries and its main sub-catchments. Pink color represents the Congo River basin and the yellow color represents the Nile River basin.

The main rivers flowing to the Nile Basin are Mukungwa, Rukarara, Base, Nyabarongo and the Akanyaru of which the water are drained by the Nyabarongo which becomes the Kagera at the outlet of the Lake Rweru (see Fig. 2.1). These important catchments were studied in this thesis except Rukarara and Base catchments due to the lack of data (see Fig. 2.2 and Chap. 3). The Kagera River forms much of the boundary between Rwanda, Burundi, and Tanzania, which drains most of the Rwandan waters to the Lake Victoria via Uganda and then to Nile River (REMA, 2005). Mutabazi *et al.* (2004) noted that Kagera River (ca. 400 km long) contributes 9 to 10% of the total Nile Waters. It is the largest inflow into Lake Victoria (24%[1] equivalent to some 7.5 km^3 of water per year), the second largest freshwater lake in the World (GEF, 2006). Therefore, Lake Victoria is of large regional importance regarding water resources.

More than 90% of the population in Rwanda depends on agriculture for their livelihood (Davis *et al.*, 2010). The production is characterised by a diversity of food crops. Most of the basin has become intensively cultivated resulting in erosion and high river sediment load from the high rainfall areas. The most important crops are bananas, beans, (sweet) potatoes, sorghum, cassava, maize, and rice (FAO, 2010). The landscape is characterized by highlands in the central and eastern part of the country which makes Rwanda to be known as "the land of the thousand hills" (Uvin, 1998).

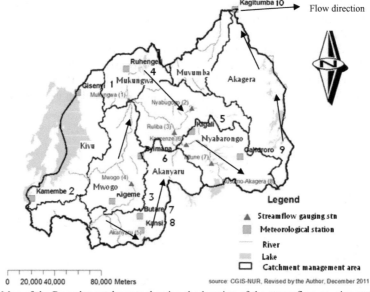

Figure 2.2 Map of the Rwandan catchments showing the location of the streamflow gauging stations and the climatic stations used in this thesis in Chapter 3.

Rwanda is characterized by a tropical temperate climate due to its high altitude. It is dominated by two wet and two dry seasons (Verdoodt and Van Ranst, 2003). The short wet season (Umuhindo) occurs from September until November (SON). The main rainy season (Itumba) lasts from March until the end of May (MAM). Between these rainy seasons two dry periods occur, a short one (Urugaryi) between December and February (DJF) and a heavy one (Icyi) from June to August (JJA).

---

[1] Or 30% of the total Lake Victoria inflow, if lake surface rainfall is included.

The amount of rainfall decreases in general from west to east (Mutabazi *et al.*, 2004). However, the annual rainfall is less than 1,000 mm a^{-1} over most of the eastern half of the country, but rises to over 1,800 mm a^{-1} in the mountainous region in the west (Mutabazi *et al.*, 2004). The temperature regime is more or less constant with a mean temperature of 16-17°C for the high altitude region, 18-21°C for the central plateau region and 20-24°C for the eastern plateau and lowlands (Twagiramungu, 2006).

2.2.2 The Migina catchment

Detailed studies were carried out in the meso-scale Migina catchment (257.4 km^2), which is located in southern Rwanda (Fig. 2.4). Approximately 103000 inhabitants with an annual growth rate of about 3% are living in the Migina catchment (Nahayo *et al.*, 2010). The geology of the Migina catchment consists of very old granite rocks, overlain by substrates of grey quartzites and schists. These geological differences result in differences in topography. The site is mountainous with elevation ranging from 1,375 m a.s.l. at the outlet to 2,278 m a.s.l. at Mount Huye, which is located in the north-western part of the Migina catchment. Table 2.1 summarizes the main characteristics of the five sub-catchments.

Migina is the name of the perennial river until it flows into the Akanyaru River, which forms the border between Rwanda and Burundi. The Akanyaru River drains into the Kagera River, which in turn flows into Lake Victoria and later generates the White Nile. Based on measurements carried out between 1971 and 1988, the discharge of the Akanyaru river is 21 m^3 s^{-1} on average (Pajunen, 1996). The main flow direction in the catchment is from north to south. The main stream is located in the eastern part of the catchment. Therefore, most of the valleys drain from north-west to south-east towards the main stream.

In the whole Migina catchment, open springs are present (see Fig. 2.14). Most of them are located at the contact between the hillslope and the valley. The springs flow throughout the year, also in the dry season (Van den Berg and Bolt, 2010).

Table 2.1 Migina catchment and sub-catchments characteristics.

Sub-catchment name (code)	Catch. area (km^2)	Total Rainfall (mm a^{-1})	Basin slopes (%)	Impervio usness (%)	Land use (%)			
					Agricult ure	Forests	Grass/L awn	Urban areas
Munyazi (W380)	38.6	1453.0	15.8	3.5	90.2	8.2	0.0	1.6
Mukura (W410)	41.6	1665.5	19.5	2.8	84.9	11.5	1.4	2.2
Cyihene-Kansi (W400)	69.6	1456.6	12.5	6.3	89.4	5.8	0.0	4.8
Akagera (W650)	32.2	1507.0	20.8	8.5	87.9	12.1	0.0	0.0
Migina outlet (W640)	61.1	1415.2	18.6	4.5	100.0	0.0	0.0	0.0

The topographic conditions are very variable and slopes of the valleys vary from 5 to 10% in the upstream and 1 to 15% in the downstream part (average slope of the sub-catchments varies between 2 and 3%) (see Table 2.1 and Nahayo *et al.*, 2010). The soils in the valleys are often ferrallitic with a 50 cm thick humic A-horizon, which are sometimes buried below dynamically colluviating deposits (Van den Berg and Bolt, 2010). The clay content of the A-horizon varies between 12% and 19% with hydraulic conductivities estimated between 1 and 10 m d^{-1} (Moeyersons, 1991).

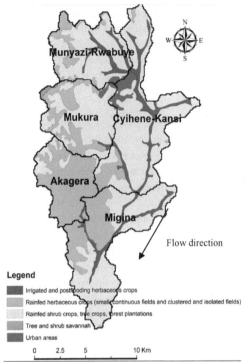

Figure 2.3 Land use of Migina catchment and sub-catchments (Munyaneza *et al.*, 2011b; adapted).

As depicted in Figure 2.3, the land cover/landuse in the Migina catchment is dominated by agricultural activities (91.2%). Forests occupy 6.5%; grass/lawn areas 0.2%, and urban areas 2.0% only. This land use distribution indicates that most of the water in the Migina catchment is used for agricultural purposes (rain-fed or irrigation).The catchment boundaries were delineated from the Digital Elevation Model (DEM) map obtained from the USGS website[2] with a resolution of 90 m using GIS tools and sub-catchment areas were generated automatically by HEC-GeoHMS 5.0 with ArcGIS 10.0. The catchment was subdivided into 5 sub-catchments as shown in Figure 2.4. Two sub-catchments are located upstream; Munyazi-Rwabuye (38.6 km²) and Mukura (41.6 km²); two in the center, Akagera (32.2 km²) and Cyihene-Kansi (69.6 km²); and one, which also contains the outlet of the whole catchment: Migina (61.1 km²) (see Table 2.1).

---

[2] http://www.dgadv.com/srtm30/

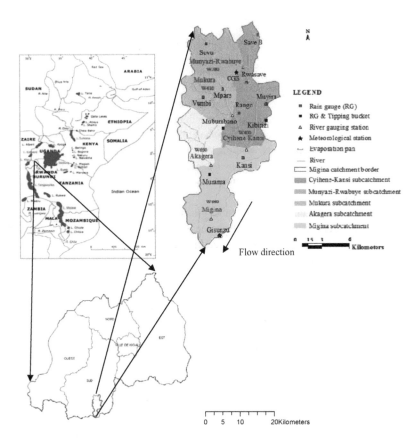

Figure 2.4 Location of the Migina catchment in Rwanda and East Africa, and instrumentation set-up within this research project (Munyaneza *et al.*, 2012a; adapted).

The Migina catchment has a moderate climate with relatively high rainfall and an annual cycle of two rainy seasons, March to May and September to November (FAO, 2005). The mean annual rainfall in the Migina catchment is approximately 1,200 mm a^{-1} and the mean annual temperature is about 20°C (SHER, 2003). The annual average evaporation in the area is estimated to 917 mm a^{-1} (Nahayo *et al.*, 2010). The mean relative soil moisture calculated for over 11 years is 75.7% with minima of 59.8% in June and the maximum in of 86.3% April (Nahayo *et al.*, 2010). The rainfall events occurred during the Itumba season (March to May) for the years 2010 and 2011 were investigated in Chapter 4.

2.3. INSTRUMENTATION

2.3.1 Meteorology

During this research historical hydro-climatic data availability was reviewed and hydro-climatic variability in Rwanda was analyzed (Chap. 3). This chapter describes the setup of hydrological and meteorological stations in the Migina catchment. The rainfall (P) as main input of the catchment, and the output of water towards the atmosphere, the total evaporation (E), was measured. The following sections discuss in details the methodology used in this research timeframe.

2.3.1.1. Rainfall

In order to obtain proper quantitative measurements of rainfall, eleven manual rain gauges have been installed throughout the catchment in the period of 24th April until 19th May at Primary schools: Save, Sovu, Mpare, Vumbi, Muyira, Rango, Kibilizi, Mubumbano, Kansi, and Murama. The twelfth was already installed at the UR fishpond of Rwasave and the thirteenth rain gauge was installed on 10th June, connected to the automated weather station at Gisunzu Primary school (see Fig. 2.5). The locations of these rain gauges were initially selected in such a way that a proportional distributed network of rain gauges was installed to cover the whole of Migina catchment. All rain gauge stations were installed near Primary Schools and were made from cheap materials (see Fig. 2.5) for security reasons and for getting accurate readings done by teachers. Local people were trained on how to carry out data collection and on the important benefits of the installation and setup of hydro-meteorological stations in the catchment.

Rainfall stations were built using PVC tubes grounded with concrete. The height is 50 cm to 1 m; on average around 70 cm above the ground surface (Fig. 2.5). Funnels and plastic cylinders were used for some stations and manual rain gauges were used at other stations. Daily rainfall data were collected at 7AM using trained local people. Rainfall intensity was measured using 5 tipping buckets installed in the catchment (see example in Fig. 2.5 left side). Data used in this research are for 2 years from May 2009 to June 2011.

Figure 2.5 Meteorological station (right side) and manual (center) and automatic (tipping bucket) (left sides) rainfall gauges installed in the Migina catchment.

To get a better understanding of the structure and evolution of single rainfall events, a tipping bucket[3] was installed next to the rain gauges of Kibilizi, Mubumbano and Murama. Unfortunately, the tipping bucket installed at Kibilizi stopped working (Fig. 2.7). A fourth tipping bucket was installed at the meteorological station at Gisunzu station. A fifth tipping bucket was already installed at CGIS meteorological station at Butare.

Besides the new data collected, historical rainfall data from all Rwandan catchments were also collected in order to review the hydro-climatic data availability and analysis of hydro-climatic variability in Rwanda (see details in chapter 3).

2.3.1.2. Spatial and temporal rainfall variability

The analysis of the spatial and temporal variability of rainfall in the whole of Rwanda was also an integral part of this research. This study presents the assessment of temporal variability of rainfall data in Rwanda primarily based on the pre-1994 period data. Daily rainfall data obtained from 136 stations countrywide were analysed in order to assess the data availability (Verdoodt and van Ranst, 2003) in many Rwandan catchments (Figs. 1.1 and 1.2). Based on these available stations, a manual screening method was used according to Munyaneza *et al.* (2009b) and finally 27 meteorological stations and 16 hydrological stations were selected for further analysis (Munyaneza *et al.*, 2010). In this research the trends and correlation analysis were carried out for historical data sets. The data sets cover a period of

---

[3] HOBO RG3 Data Logging Rain Gauge:
http://www.tempcon.co.uk/html/D233%20HOBO%20RG3%20&%20RG3-M%20Rain%20Gauges.pdf

1910 to 2008 for rainfall. Rainfall data have been recorded on a daily basis. Figure 2.6 shows inter-annual variability of rainfall, based on these selected stations.

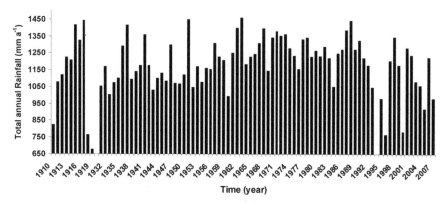

Figure 2.6 Inter-annual variability of mean rainfall based on selected stations in all Rwandan catchments (1910–2008) (Source: Munyaneza *et al.*, 2010). Stations were arranged based on their long time series availability with few missing historical data sets.

Figure 2.6 shows that there is a data gap from 1920 to 1930 and in 1994. The figure also shows the maximum annual rainfall in Rwanda from 1910 to 2008 which is 1450 mm a^{-1} in 1951 and the minimum of 677 mm a^{-1} in 1919. The years of 1918, 1919, 1995, 1996, 2000 and 2005, present drought periods in Rwanda.

Based to the national gaps observed in historical rainfall data which were also particularly observed at Save rainfall station (located in the Migina catchment), thirteen new manual rain gauges have been installed throughout the Migina catchment with five automatic rain collectors (tipping buckets). The tipping bucket data were converted to daily rainfall series (van den Berg and Bolt, 2010). These data were completed with daily rainfall measurements from thirteen manual rain gauges installed throughout the catchment (Munyaneza *et al.*, 2012a). The area covered by the rain gauges is about 260 km^2, giving a network density of slightly less than 1 per 20 km^2.

The impact of the topography on the average daily rainfall was assessed with linear regression (Chap. 5). The average daily rainfall was split per month and the monthly data were correlated with altitude, aspect, slope and east and north coordinates. Thiessen polygon method (Fig. 2.7) was used for new rainfall data collected from May 2009 to June 2011 for the interpolation of the daily rainfall average. This information helped to get an idea of the quantity of water entering Migina catchment in recent time. The method is chosen because it represents two ends in a spectrum of interpolation methods (Buytaert *et al.*, 2006). Thiessen method is a simple and straightforward method. Each interpolated location is given the value of the measurement point within the polygon, resulting in a typical polygonal variation and discontinuities at the borders of the polygons (van den Berg and Bolt, 2010). In this method, the results of the multiple regression results are incorporated (Chap. 3). The data were normalized, based on the relation of the total daily rainfall with the topographical parameters.

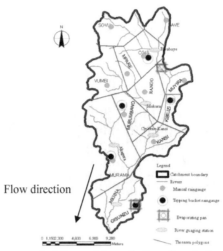

Figure 2.7 Thiessen Polygon method for spatial distribution of newly installed rain gauges stations in the Migina catchment.

2.3.1.3. Evaporation

To get an estimate about the amount of evaporation (E, expressed in mm d^{-1}) in Migina catchment, two evaporation pans, which give a measure of the pan evaporation (Epan) have been installed (see Fig. 2.8). Evaporation pans provide a measurement of the combined effect of temperature, humidity, wind speed and sunshine on the E_{pan}. Two self-made replications of the U.S. Weather Bureau 'Class A' evaporation pans were made locally, of which one was installed at the Rwasave fishpond and one at the meteorological station of Gisunzu (see Fig. 2.7 for their locations).

Figure 2.8 Evaporation pans installed at Gisunzu Primary school (right side) and Rwasave fishpond (left side) in the Migina catchment during this reseach.

Figure 2.8 shows iron pans which were made in circular shape, with a diameter of 120 cm and a depth of 25 cm as recommended by U.S. Weather Bureau, and were filled to 5 cm from the pan top (see Fig. 2.8 on right side, red color sign). The pan rests on a wooden base, to be able to install the pan carefully leveled and to minimize the effects of ground heat. The scale bar inside the pan was read out manually twice a day (at 7AM and 7PM) to determine the change in water level and thus the amount of pan evaporation (E_{pan}) presumably taking into account rainfall. Recorded data are available for 2 years from May 2009 to June 2011 for both locations, and results are presented in Chapter 5.

2.3.2 River water

For qualitative and quantitative description of surface water flows in the Migina catchment, hydrological and hydrochemical measurements have been carried out. To investigate the outflow of surface water, in total five gauging stations have been installed in the main 5 rivers at the outlet of each sub-catchment of the Migina catchment. The selection of these locations was sought, based on the criteria described by Waterloo *et al.* (2007). PVC tubes were used and iron stills were fixed with concrete for protection (Fig. 2.9). A pressure transducer (mini diver, Van Essen Instruments) was installed inside the PVC tube for automatic water level measurements (Fig. 2.10). The water pressure was measured automatically with a D or TD-Diver[4] at an interval of 15 minutes. This diver was constructed at the deepest point of the L-shaped PVC-tube and data was read out and stored on a laptop once a month (Fig. 2.10). The water pressure measurements of the divers have been corrected for atmospheric pressure variations. Therefore, a Baro-Diver[5] was installed in Migina catchment of which the measured daily variation in atmospheric pressure has been subtracted from the water pressure measurements of the divers.

Manual daily water level measurements were collected by local people twice a day at 7AM and 7PM as a back-up for the diver data. The scale of the staff gauge was made by measuring tape, which was fixed on a wooden post with nails. The post was put in a big block of concrete (>200 kg) to keep the staff gauge stable, straight and immovable in the river (see Fig. 2.9).

Manual corrections have been made for outliers in the data, which could be owing to for example, sudden shifts in water pressure caused by a change in installation depth of the diver (Fig. 2.9). After these corrections, the automatic measured water level variations have been compared with the manual staff gauge readings. The results can be found in Munyaneza *et al.* (2010) and in Chapter 5.

Figure 2.9 Installation of the L-shaped PVC-tube (upper left) and the staff gauge (lower right). The open tube with the diver inside is closed with a locked metal casing, which is fixed to concrete piles (centre). The old rehabilitated Rwabuye gauging station is shown in the upper right side. The staff gauge type as used in all five gauging stations is shown in detail in the lower right side photo.

---

[4] TD1/TD2/D2 Diver - former Van Essen Instruments (nowadays: Schlumberger Water Services)

[5] Baro Diver – former Van Essen Instruments (nowadays: Schlumberger Water Services)

To provide reference measurements of the river stage, the level of the staff gauge was surveyed relative to a local benchmark, by using a theodolite[6] and a scale-bar (Fig. 2.10). The local benchmarks are marked points on or nearby bridges.

Figure 2.10 Akagera River discharge measurement (picture taken on 10/08/2009 by Emille (BSc student) when UR Civil Engineering students were instructed on how to measure discharge using a propeller (left). Data read out (center) and staff gauge leveling (right) (Van den Berg and Bolt, 2010).

The Figure 2.10 shows how river discharge measurements were taken continuously at different gauging stations by current meter (propeller) measurements using the area-velocity method (Waterloo *et al.*, 2007). To create a velocity profile of the river, velocities were measured every 20 cm along a cross-sectional profile of the river. At these locations the velocity of the water was measured at 60% of the channel depth.

The frequency of river discharge measurements during this research were based on change in water level (high flow and baseflow) (Waterloo *et al.*, 2007), to establish a water level-discharge relationship (see in Munyaneza *et al.*, 2010). This relationship so called rating curve (Eq. 2.1), generated referring to the recommendations of Shaw (2004), is used to make time series of continuous river discharge called hydrographs as described in van Breukelen *et al.* (2008). With these hydrographs, better understanding of hydrological processes was gained into runoff generation processes and surface water-groundwater interactions (Chap. 4).

The rating curve is often represented by an equation of the form (De Laat and Savenije, 2002; De Laat, 2006):

$$Q = a(H - H_0)^b \tag{2.1}$$

where Q is the discharge in $m^3 \, s^{-1}$, H is the water level in the river in m, H_0 is the gauge reading at zero level, and *a* and *b* are constants. The value of H_0 is determined by trial and error while the values *a* and *b* are found by a plot on logarithmic paper and fit of a straight line or by a least square fit using the measured data.

Equation (2.1) is compatible with the Manning formula (Eq. 2.2) where the cross-sectional area, *A*, and the hydraulic radius, *R*, are functions of (H-H_0).

$$Q = \frac{A}{n} R^{\frac{2}{3}} S^{\frac{1}{2}} \tag{2.2}$$

where *n* is the Manning roughness coefficient [-] and *S* is the slope of the channel (dimensionless).

---

[6] Sokkia – Electronic Digital Theodolite DTX20 series - DT520:
     http://www.sokkia.com/Products/Detail/DT620.aspx

2.3.3 Groundwater

To investigate the level changes of the groundwater and its relationship with runoff generation, two transects with eleven piezometers were installed in Kadahokwa marshland located in west-central of Migina catchment (Fig. 2.11), and continuous water level measurements were carried out and groundwater samples were collected and analyzed. The detailed results can be found in Chapter 4.

The piezometers are placed in two parallel transects, between the hillslope and river. The distance between the two transects is approximately 60 m. Transect 'A' consists of 7 piezometers, transect 'B' consists of 4 piezometers. An overview of the locations of the piezometers is given in Figure 2.11.

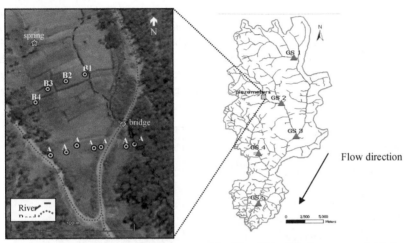

Figure 2.11 Locations of the piezometer transects A1 - A7 and B1 – B4, shown on an aerial photograph (left, © Google Earth) and on a map (right) (source: van den Berg and Bolt, 2011).

The piezometers are constructed from polyvinyl chloride (PVC) pipes. The bottoms of the tubes are closed with concrete to avoid mud entering inside. The tubes are provided with a screen of small slots over a length that varies between 0.3 and 1.6 m. The screen is made by perforating the PVC tube with a saw. The length of the screen is chosen depending mainly on soil material and the thickness of permeable layers that are found. If possible, the screen is located at a depth where permeable layers are present. The permeable layers that are found contain sands and gravels. Before installing, filter stocking is put over the slots to prevent the tube from sedimentation and clogging.

After preparation, the tube is placed in the drilled hole. Then the borehole (along the tube) is filled up. First, next to the screen, the borehole is filled up with a gravel pack, to prevent the filter from clogging. On top of these gravels, the original material is replaced. Finally, on top of the tubes metal casings are placed, that can be closed with a lock door for the diver security and the casings are fixed in concrete (see Fig. 2.12).

Figure 2.12 Piezometers installed in Kadahokwa marshland (West-Central of Migina Catchment).

In accordance to statistical data that was obtained by Kasanziki (2008), 8% of the people in the area around Butare use piped water as their main water source and 81% of the people use springs as main source of water (Fig. 2.13). The remaining 11% relies on water from rivers (5%), lakes/ponds (5%) and other sources such as water harvesting, etc (1%). The public company in charge of water supply (called EWSA) is only able to deliver its treated water to Butare city (located upstream of the catchment) and some of their urbanized areas nearby (van den Berg *et al.*, 2010). The springs are mostly located at the foot of the hills. In most cases, the spring is made of a pipe, which is connected to the groundwater table (Fig. 2.13).

Figure 2.13 Open springs in the Migina catchment used by local people to collect drinking water.

The water that comes from the springs is shallow groundwater, mostly located at the contact between the hillslope and the valley floor. In the Migina catchment, more than fifty springs might be present (Kasanziki, 2008). This number is a rough estimation, as precise numbers and locations are not available. During this study, a selected number of the springs are located.

Twelve springs are located in the northern part ((1) sovu, 1733 m, (2) Gahenerezo 1671 m; (3) Nabagabo, 1652 m; (4) Kinteko, 1643 m; (5) Rwasave, 1665 m; (6) Mpare, 1652 m; (7) Kamugenge, 1682 m; (9) Gakombe, 1694 m; (10) Kadahokwa, 1646 m; (11) Mpazi, 1633m; (12) Rango (1614m; and (13) Kibingo, 1575 m; all m.a.s.l.), and only two in the south ((14) Nyamugali, 1536 m; (15) Nyarunazi, 1542 m; all m.a.s.l.) (see Fig. 2.14 and Table 2.2 their locations). It is expected that some more springs can be found in the north, and much more in the south (van den Berg and Bolt, 2010). The results of the water sampled and analyzed for its hydrochemical and isotope contents can be found in Chapter 4.

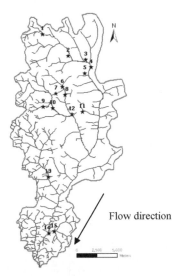

Flow direction

Figure 2.14 Locations of investigated springs in the Migina catchment.

A handheld GPS receiver[7] was used to measure the UTM coordinates of the locations. This GPS receiver has an indicated error for the x and y coordinates of approximately 3 m. UTM coordinates are shown in Table 2.2.

Table 2.2 Investigated springs in the Migina catchment with UTM coordinates.

#	Name	UTM coordinates		
		X	Y	Z
1	Sovu	800948	9717210	1733
2	Ghahenerezo	804136	9714635	1671
3	Nabagado	806274	9714124	1652
4	Kinteko	806973	9713221	1643
5	Rwasave	806184	9712510	1665
6	Mpare	803405	9710842	1652
7	Kamugegemge	802604	9710041	1682
8	Runyinya	803703	9709864	1644
9	Gakombe	801008	9708431	1694
10	Kadahokwa	802179	9708263	1646
11	Mpazi	805879	9707830	1633
12	Irango	804610	9707503	1614
13	Kibingo	801597	9699935	1575
14	Nyamugali	801646	9693089	1536
15	Nyarunazi	802291	9693333	1542

2.3.4 Hydrochemistry

To gain better estimate into the different sources and pathways of the surface water and groundwater, hydrochemical field analyses as well as sampling for hydrochemical laboratory analysis have been

---

[7] Garmin eTrex Vista HCx
http://www.garmin.nl/product/?pid=010-00630-01

executed (Fig. 2.15). Hydrochemical and isotope data were collected over two years (May 2009 to June 2011). The collected samples include groundwater from 11 shallow piezometers, 15 springs, streams water sampled at 8 sites in the catchment (weekly or monthly intervals), and monthly catchment rainfall from 5 rain collectors (where tipping buckets installed).

After the construction of the piezometers, the water in the tubes is removed by using a hand pump. Then, the tube is filled with water from the river. This procedure was carried out at least three times, to be certain that the water from the piezometers was not influenced by the materials that were used to build the piezometer.

After a week, the water in the piezometer was removed with the pump, to get fresh water in the tube. This way, the influence of the introduced river water is supposed to be negligible. The water in the tubes is now supposed to be the original water and ready to be sampled. Sampling from piezometers took place after refreshing the volume of the piezometers at least three times with use of a hand-pulse pump (Fig. 2.15 on left side).

Figure 2.15 Water sampling from piezometers and lab work for water chemistry analysis.

The groundwater levels in the tubes are measured continuously, using a deeper (water level sounding meter) (Fig. 2.15 left side). Pressure transducers (Mini-Diver[8]) are installed to obtain automatic water level and temperature records every 15 minutes. In one of the piezometers a Baro-Diver[9] is installed to obtain barometric data. The data of the Baro-Diver is used to correct the groundwater level measurements for changes in air pressure. Rainfall event for Itumba'11 season was also collected at Gisunzu rain gauge for isotopic composition analysis and the results are presented in Chapter 4.

In this current study, in-situ measurements have been continuously conducted for pH, water temperature (T) and electrical conductivity (EC) using an EC-pH meter. Stream, piezometer, spring and rain water samples were collected in 30 ml plastic bottles which were filled as much as possible and closed with special cups to prevent evaporation from the bottles. Samples were collected during both low flows and flood events and analyzed in the laboratory. The details of the methods used for laboratory analysis and results are shown in Chapter 4.

2.3.5 Satellite imagery
In order to have a good picture of the study area, Digital Elevation Model (DEM) maps obtained from the USGS website[10] with 90 m resolution and remote sensing data were needed and collected from CGIS-UR. These data helped to get information related to agricultural practices and statistics (e.g. crops and yields), land use and land cover in the catchment for the assessment of water resources availability in the Migina catchment. Various criteria were applied in selecting the images (Munyaneza *et al.,* 2009a). Suitability of images and their availabilities refer to the data quality in relation to the proposed application. The results are available in Chapters 5 and 6 of this thesis.

---

[8] Schlumberger Water Services
http://www.swstechnology.com/groundwater-monitoring/groundwater-dataloggers/mini-diver

[9] Schlumberger Water Services
http://www.swstechnology.com/groundwater-monitoring/groundwater-dataloggers/baro-diver
[10] http://www.dgadv.com/srtm30/

Chapter 3

STREAMFLOW TRENDS AND CLIMATE LINKAGES IN MESO-SCALE CATCHMENTS IN RWANDA

....

A study on streamflow trends and climate linkages in four meso-scale catchments in Rwanda was conducted over a long period of time more than 30 years (between 1961 to 2000). Streamflow records from major catchments contributing to the Kagera River, which is the main inflow to Lake Victoria, were studied on seasonal and annual scales using a number of hydrological indices. Rainfall and air temperature data from 10 meteorological stations, ranging from 30 to 59 years of data between 1935 and 2008, were used to detect trends in climatic variables and their correlations with streamflow data of selected meso-scale catchments. Trends and the time of change points were investigated using the Mann-Kendall test and Pettitt test, respectively. The significance of the change of the two sub-series has been tested separately using a t-test at significance level of 5 %. The linkages between each of the climatic and hydrological variables were investigated using Pearson correlation. The results revealed significant trends for climatic and hydrological variables and an overall increasing trend of streamflow was detected in the longer period of data (1961-2000) and a decreasing trend in the short period of the study (1971-2000). Pettitt tests revealed that abrupt change points of most stations occurred in the 1980s–1990s, which is related to the period of intensive human activities in Rwanda such as agriculture development and urbanization.

....

Based on: Munyaneza, O., Twahirwa, A., Maskey, S., Uhlenbrook, S. and Wenninger, J., 2012. *Streamflow trends and climate linkages in meso-scale catchments in Rwanda.* IWA Publishing, submitted.

3.1. INTRODUCTION

Analyzing time series for trend detection of hydro-meteorological variables has received a great deal of attention over the past decades because of its importance for understanding and predicting future climatic change (e.g. IPCC, 2007). Some of the detected trends could be attributed to climatic (i.e. rainfall and temperature), which may lead to altered flood or low flow regimes that affect water resources and infrastructures (e.g. Uhlenbrook, 2009). Sharif and Burn (2009) demonstrate that operational decisions for water resources infrastructure and management are dependent on both the timing and magnitude of flows and, therefore, climate change impact assessment should consider both types of characteristics. Some recent studies that investigated linkages between climate indices, precipitation and streamflow are Burn and Cunderlik (2004), Chingombe *et al.* (2005), Sharif and Burn, (2009), Love *et al.* (2010), Masih *et al.* (2010), and Hu *et al.* (2011a and b).

A number of studies have examined trends of river discharge at scales ranging from headwater catchments to river basin to continental/global scales. IPCC (2007) reviews the synthesized results of many studies published before 2007. IPCC (2013) includes many more recent studies that largely confirm the main results and provide many additional insights. Mikova *et al.* (2009) predict lower agricultural production and food security in many African countries and regions due to the result of global warming. It is expected that size of arable land, the duration of vegetation period and crop yield will reduce in East Africa (Molden *et al.*, 2007). In this regards, REMA (2009) analyzed rainfall over the last 30 years, and found in some parts of Rwanda irregularities in climate variation including variability in rainfall frequencies and intensities. This study of REMA (2009) shows that rainy seasons in Rwanda are tending to become shorter with more intensive rainfall events. Consequently, this tendency has led to a decrease in agricultural production and more frequent droughts as reported by Mikova *et al.* (2009). On the other hand, floods and landslides are experienced in areas with heavy rains (REMA, 2009). However, these studies did not include trends in the hydrological regimes (Hu *et al.*, 2011b). Furthermore, streamflow trends and their linkages with climate are not well understood in Rwanda. Therefore, this study attempts to fill an important knowledge gap and is mainly focused on the trends on streamflow and the linkage between climate and streamflow trends (Masih *et al.*, 2010).

A study on streamflow trends and climate linkages in eight meso-scale catchments in Rwanda was conducted over a period of 40 years (1961-2000). Streamflow records from major catchments contributing to the Kagera River were studied on seasonal and annual scales using a number of hydrological indices.

3.2. DATA AND METHODS

3.2.1. Selection of stations

Data were obtained from the Ministry of Natural Resources (MINIRENA) for eight streams (Table 3.1), and from the Rwanda Meteorological Service under the Ministry of Infrastructure (MININFRA) for climate data (Table 3.2). The minimum length of streamflow record is 40 years (1961-2000) at St3 and St8 and 30 years (1971-2000) for the remaining stations. In principle, only four stations (St3, St5, St6 and St8) can be used for analysis, but only after data correction and also only for a limited time due to large data gaps (Table 3.1). Hence, they were finally adopted in the current study to detect trends in streamflow variables due their modified rating curves were updated and corrected by RIWSP Programme (RIWSP, 2012a and 2012c). The temporal resolution of data available is daily. Station St1, St2, St4, and St7 cannot be used for the analysis because of unreliable discharge data. The reasons are mainly due to: i) gauge datum shifts and gradual shifts in time of zero stage control levels, ii) poor river bed stability, iii) instable controls; and iv) insufficient number of discharge measurements. The general conclusion which can be drawn after the analysis is that there is no continuous time series of daily water levels and/or discharge data available in Rwanda for the selected time period. All assessed gauging stations show a gap between at least 1990 and 1995. The percentage value of missing streamflow data varies between 29 and 46% in the daily time series (see Table 3.1).

Similar to streamflow stations, climate (rainfall and temperature) stations were selected based on long time series data availability and few missing data observed in the records (see Table 3.2). This resulted in 10 climate stations over the four different periods (1935-1993, 1960-1994, 1971-2000 and 1961-2008) were adopted in the current study to detect trends in climatic variables and their correlations with the streamflow data (Masih, 2011). The areal rainfall was estimated using the Thiessen polygon method from the 10 stations within the 136 stations collected from Meteo Rwanda database (SHER, 2004).

The data used in this study were first processed by Meteo Rwanda (2002) and Mikova *et al.* (2010) for climate data, and by SHER (2004) for both streamflow and climate data. The data quality control and homogeneity tests were also examined in this study on data from selected stations using the double mass curve method as suggested by Mann (1945) and Kendall (1975). In case of missing data as shown in Tables 3.1 and 3.2, the time series have been interpolated using linear interpolation method by the IHA software (Masih *et al.*, 2010). Richter *et al.* (1997) discussed this issue, as well as various methods for extending hydrologic records, filling in missing data, or estimating daily hydrologic data from simulation modelling.

The IHA software uses daily data for its calculations. The IHA statistics are meaningful only when calculated for a sufficiently long hydrologic record. The recommended length is at least twenty years of daily records for trend analysis (IHA, 2001). Interpolated values will be generated for all days with missing data that are in water years that have at least one valid flow value (Richter *et al.*, 2009). For IHA version 7.1 used in this research, linear interpolation is excluding from analysis any water year with no valid flow data. Note that in years with very large gaps in the datasets, many interpolated values can lead to odd results for rise/fall rates, pulses, and other parameters. It is in this regards, data collected from MINIRENA which have such problems were not used in the research. However, hydro-climatic regimes in Rwanda which were used in this research can be found in Figure 3.1.

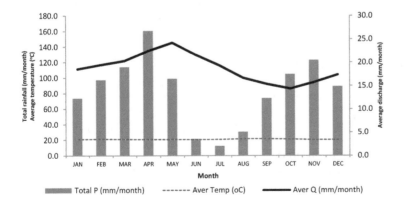

Figure 3.1 Hydro-climatic regimes in Rwanda. The rainfall and temperature data was used from Kigali Airport station (1961-2008), the discharge data presented originates from Rusumo-Kagera station (1961-2000).

Figure 3.1 shows the hydro-climatic regimes in Rwanda. The maximum monthly total rainfall is observed in April (160 mm month[-1]) for data collected at Kigali Airport station (1961-2008). The maximum average discharge is observed in May (286.5 m³ s[-1] equal to 24 mm month[-1]) for data collected at Rusumo-Kagera station (1961-2000). This means that the increasing or decreasing rainfall in April might cause increasing or decreasing May flow at Kagera river (St8). The maximum monthly temperature is observed in August (21.6°C), which is in the dry season in Rwanda and affects the minimum monthly flows observed the next month in September (118.1 m³ s[-1] equal to 15 mm month[-1]) at Kagera river. Temperature affects flow directly in such a way that temperature can be seen as a proxy for the atmospheric evaporation demand.

The selected streamflow stations within the Nile River basin are: Akanyaru river (Butare-Ngozi road); Nyabarongo river (Ruliba, Kanzenze, Mfune); Mwogo river (Nyabisindu); Mukungwa river (Ngaru-Kageyo), Nyabugogo river (Cyamutara); and Kagera river (Rusumo). The climate stations are well distributed in these meso-scale Rwandan catchments (Fig. 2.2).

Table 3.1 Description of the stream flow gauging stations. The entries in bold indicate stations which cannot be used for the analysis because of unreliable discharge data.

Station name (code)	River name	Lat	Long	Altitude	Upstream area	Rainfall	Discharge	Record Length	Missing data
		(°N)	(°E)	(m asl)	(km²)	(mm a⁻¹)	(mm a⁻¹)		(%)
Ngaru-Kageyo (St1)	**Mukungwa**	**-1.73**	**29.65**	**1400**	**2100**	**1381**	**433**	**1971-2000**	**39**
Cyamutara (St2)	**Nyabugogo**	**-1.79**	**30.13**	**1439**	**1647**	**996**	**201**	**1971-2000**	**41**
Ruliba (St3)	Nyabarongo	-1.97	30.00	1352	8336	1000	376	1961-2000	31
Nyabisindu (St4)	**Mwogo**	**-2.35**	**29.67**	**1525**	**534**	**1230**	**289**	**1971-2000**	**42**
Butare-Ngozi (St5)	Akanyaru	-2.81	29.82	1406	1491	1306	278	1971-2000	29
Kanzenze (St6)	Nyabarongo	-2.06	30.09	1337	13930	1094	271	1971-2000	45
Mfune-Kibungo (St7)	**Nyabarongo**	**-2.20**	**30.28**	**1337**	**15547**	**1158**	**261**	**1971-2000**	**34**
Rusumo (St8)	Kagera	-2.38	30.78	1325	30644	906	227	1961-2000	32

Table 3.2 Description of the temperature and rainfall stations. The temporal resolution is daily.

Station name (code)	Lat	Long	Altitude	Temperature			Rainfall		
				Mean Annual	Record	Missing	Mean annual	Record	Missing
	(°N)	(°E)	(m asl)	(°C)	Length	(%)	(mm a⁻¹)	Length	(%)
Gisenyi (Pt1)	-1.67	29.25	1554	20.14	1975-2000	30.0	1178	1971-2000	30.4
Kamembe (Pt2)	-2.47	28.92	1591	20.15	1971-2000	15.4	1383	1971-2000	20.8
Kigeme (Pt3)	-2.22	29.12	2124	17.57	1960-1994	25.0	1464	1960-1994	1.7
Ruhengeri (Pt4)	-1.50	29.60	1878	17.65	1977-1994	13.3	1405	1960-1994	10.0
Kigali (Pt5)	-1.97	30.13	1490	20.95	1971-2006	1.9	1000	1961-2008	2.0
Byimana (Pt6)	-2.18	29.73	1750	18.24	1960-1994	2.9	1120	1960-1994	6.6
Butare (Pt7)	-2.60	29.73	1760	21.79	1935-1993	0.0	1251	1935-1993	2.9
Kansi (Pt8)	-2.72	29.75	1650	21.76	1981-1993	0.0	1158	1935-1993	1.9
Gahororo (Pt9)	-2.17	30.50	1700	19.27	1960-1994	2.9	1159	1960-1994	1.7
Kagitumba (Pt10)	-1.05	30.43	1280	19.52	1956-1980	0.0	754	1971-2000	3.4

3.2.2. Selection of hydrologic variables

Hydrological variables can serve as important indicators of climatic change. Many previous studies have suggested different variables for detecting climatic changes (e.g. Chingombe et al., 2005; Hu et al., 2011b). The timing and duration of hydrologic events were considered and a total of 11 hydrological variables (indices) were selected (Table 3.3). The procedure adopted in selecting the indices is to select a collection of variables encompassing the important components of the hydrologic regime as proposed by Chingombe et al. (2005). The software IHA version 7.1 (Indicators of Hydrologic Alteration) (Richter et al., 2009) was used to derive the indices from the daily time series.

3.2.3. Trends detection test

Different reviews of applications of parametric and non-parametric methods have been done for the detection of trends (e.g. Kundzewicz *et al.*, 2004; Chen *et al.*, 2007). Parametric tests are more powerful than non-parametric ones, but the parametric tests make an implicit assumption of normality of data that is seldom satisfied (Sharif and Burn, 2009). Hydroclimatic time series are often characterized by data that is not normally distributed, and therefore non-parametric tests are considered more robust compared with the parametric tests (Hess *et al.*, 2001). In this study we used the Mann-Kendall non-parametric test for detecting trends (see Hu *et al.*, 2011b). This test allows investigating long-term trends without assuming any particular distribution, and this test is also less influenced by outliers in the data sets (Hu *et al.*, 2011b). Several researchers have used the Mann-Kendall test to identify trends in the hydro-climatic variables with focus on climate change (e.g. Douglas *et al.*, 2000; Chen *et al.*, 2007; Burns *et al.*, 2007; Sharif and Burn, 2009). In this study, statistical significance of the trends is evaluated at the 5% level of significance against the null hypothesis that there is no trend in the analyzed variable (Table 3.3). The used test statistic S of the Mann-Kendall test can be found in Mann (1945); Kendall (1975); Hirsch *et al.*, (1982); Burn and Hag Elnur (2002); and Hu *et al.* (2011a,b).

In this study, the Mann Kendall non-parametric trend test has been performed using XLSTAT software (Shetkar and Mahesha, 2011). A positive value of the test statistic S indicates an upward and a negative value indicates a downward trend (Helsel and Hirsch, 1992).

The standard normal distribution was then used for hypothesis testing, and is called in this study the trend test statistic Z, which is calculated by:

$$Z = \begin{cases} \dfrac{S-1}{\sqrt{VAR(S)}} & \text{if } S > 0 \\ 0 & \text{if } S = 0 \\ \dfrac{S+1}{\sqrt{VAR(S)}} & \text{if } S < 0 \end{cases} \tag{3.1}$$

Where the test statistic *(S)* and variance, *VAR(S)*, are defined e.g. in Helsel and Hirsch (1992) or Burn *et al.* (2004). The standardized Z value is used to determine the significance of any trend in the data set. If $|Z| > Z_{1-\alpha/2}$, the null hypothesis is rejected at significance level α that indicates the trend strength. In this study, significance level of 5% is applied and the observed *p*-value is obtained for each analyzed time series.

3.2.4. Change point detection

The Pettitt test, which is an approximation for a sequence of random variables of the non-parametric method, has been used in this study and helps to indicate where possible change points are (P > 0.8) (Pettitt, 1979). The Pettitt test only detects the time of a change point (see Table 3.4), but a t-test was always applied to check the significance of the trend. A significance level of 5 % has been applied in this study. Further details about the use of this method can be found e.g. in Ashagrie *et al.* (2006); Zhang *et al.* (2009), and Love *et al.* (2010).

The approximate significance probability *(P)* for a change point is defined by Equation (3.2):

$$P = 1 - \rho \tag{3.2}$$

Where: P is the probability of detecting the point of change and ρ is the existence of significant change point (Zhang *et al.*, 2009).

The ρ -value (two-tailed) has been computed in this study using XLSTAT software with 10,000 Monte Carlo simulations at 99% confidence interval for checking the data homogeneity (Shetkar and Mahesha, 2011). Hirsch *et al.* (1982) showed that it is possible to obtain a non-parametric estimate for the magnitude of the slope using Eq. (3.3) to facilitate the comparison of results with different flow magnitudes,

$$\beta = Median\left\{\frac{x_j - x_k}{j - k}\right\} \qquad \text{for all } k < j \qquad (3.3)$$

where: β is a robust estimate of the slope.

3.3. RESULTS
3.3.1. Streamflow and climatic trends

3.3.1.1 Mann-Kendall test
Eight streamflow time series are tested for trends in 11 hydrological variables. A summary of the MK trend results is presented in Table 3.3.

Table 3.3 Trend test results for selected streamflow variables. The entries in bold indicate Z-values that are significant at the 5% level. Positive and negative signs of Z indicate increasing and decreasing trend, respectively.

Group	Streamflow variables	Units	St3 1961-2000	St5	St6 1971-2000	St8 1961-2000
Group 1	Annual mean flow	m^3 s^{-1}	1.92	-1.92	**-3.34**	**2.46**
	Urugaryi (DJF)	m^3 s^{-1}	1.01	-1.17	-1.8	**3**
	Itumba (MAM)	m^3 s^{-1}	1.03	-1.36	**-2.83**	1.55
	Icyi (JJA)	m^3 s^{-1}	1.48	-1.44	**-2.64**	0.91
	Umuhindo (SON)	m^3 s^{-1}	**2.55**	-1.81	-1.38	**5.07**
	1-day min	m^3 s^{-1}	**3.03**	-0.85	-1.52	**3.81**
	7-day min	m^3 s^{-1}	**3.02**	-0.83	-1.33	**3.83**
Group 2	1-day max	m^3 s^{-1}	1.18	-1.56	**-2.97**	0.49
	7-day max	m^3 s^{-1}	0.61	-1.1	**-2.78**	0.52
Group 3	Date of minimum	Julian day	1.46	-0.5	-1.28	**-2.26**
	Date of maximum	Julian day	-0.17	1.31	-0.16	-0.79

It is shown the trend analysis results for the two study periods (1971-2000 and 1961-2000) for the 11 selected hydrological variables. More significant trends are detected when the length of the study period is increasing. Station St8 presents strong increasing significant trends at the 5% significance level in almost all variables. For the shorter study period (1971-2000), St5 did not show significant trends at the 5% level for all hydrologic variables.

Table 3.3 also shows that some streamflow variables have trend results but with some particularities. Strong trends are also noted at the same group 1 for annual mean flow, DJF and SON seasonal mean flows, while slight trends were noted for MAM and JJA seasonal mean flows. For group 2, 1- and 7-days minimum flows and baseflow index were observed to have strong trends while the 1- and 7-days maximum flows show a lack of trends except at St6. From group 3, statistically significant decreasing trends were exhibited at St8 in the date of minimum while no significant trends in the date of maximum were detected at the 5% level at all stations.

There were also several significant trends that were observed only for a particular analysis period. The magnitude of the MAM and JJA seasonal flows show an opposite behavior from station St6 over 1971-2000 periods. Strong decreasing trend is apparent at station St6 (downstream).

3.3.1.2 Change point results

Pettitt test is used to identify a change point in a time series (Eq. 3.2), and assumes that the observations form an ordered sequence (Pettitt, 1979; Zhang et al., 2009).

Table 3.4 Results of the Pettitt test for selected streamflow stations. P is the probability of detecting a change of point at significance level of 5%.

Station name	Station code	River name	Change point	P Annual flow	Shift	Change point	P 1-day annual min flow	Shift
Ruliba	St3	Nyabarongo	1987	0.997	Upward	1985	1	Upward
Butare-Ngozi	St5	Akanyaru	1990	0.96	Downward	1991	0.995	Downward
Kanzeze	St6	Nyabarongo	1984	1	Downward	1981	0.974	Downward
Rusumo	St8	Kagera	1962	0.991	Upward	1968	1	Upward

Table 3.4 shows the results from Pettitt tests that reveal statistically significant shifts for annual flows and the extreme minimum flow for 1-day annual minimum flows at the significance level of 0.05. This is the lowest daily minimum flow along the Rwanda water year (January to December) for the examined stations.

It is clear from Table 3.4 that both two main rivers Akanyaru (St5) and Nyabarongo (St6), which form Kagera River (St8), faced a shift downward of annual flows in the years 1990 and 1984, respectively, with significant abrupt changes for significance level alpha equal to 0.05. Similar results can also be seen at both stations (St5 and St6) in the 1-day annual minimum flows for the years 1991 and 1981, respectively. An opposite observation is observed at St8 in 1962, which reveals a shift upward of annual flows while it is located downstream of St5 and St6 (see Fig. 2.2). However, in St5 and St6, the urbanization could be influencing more river water runoff and less groundwater recharge and therefore, annual minimum flows goes down while in other stations annual minimum flows go up which could be the results of water storage and reservoir development. For the annual flow, no clear explanation could be found at the moment.

3.3.1.3 Climatic variables

Climatic data were also investigated in the studied catchments for trends over four different periods (1935-1993, 1960-1994, 1971-2000 and 1961-2008) at 10 selected stations based on long time series data sets availability with few missing data ranging between 0.0-30.4% (Table 3.2). The available data included minimum, maximum and mean temperature as well as total rainfall. Trends were investigated on annual, seasonal and monthly basis (Table 3.5). Time series for minimum, mean and maximum temperature for the period from 1971 to 2006 at Kigali station are shown as an example in Figure 3.2.

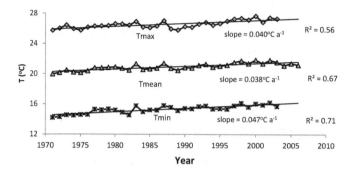

Figure 3.2 Time series plots for minimum, maximum and mean temperature from 1971 to 2006 at Kigali (Pt5) station. The symbols show the observed annual mean values and straight lines show the linear trend line. R^2 shows the coefficient of determination.

Figure 3.2 shows that the three temperature variables gave similar results in terms of increasing trends. This was observed in this study from almost all selected stations over the study period. Therefore, the summary shown in Table 3.5 only refers to the mean temperature data.

Table 3.5 Mann-Kendal non-parametric trend test results for some climatic (rainfall and temperature) indices. The entries in bold indicate values that are significant at the 10% level, due to fact that no trends were detected at 5% level at almost all stations. Positive and negative signs indicate increasing and decreasing trends, respectively. The entries in brackets represent the values (Z) for mean temperature.

Climatic	Pt1	Pt2	Pt3	Pt4	Pt5	Pt6	Pt7	Pt8	Pt9	Pt10
season	1971-2000		1960-1994		1961-2008	1960-1994	1935-1993		1960-1994	1971-2000
Annual	**-1.78** (0.88)	**-1.82** **(3.44)**	-0.58 (0.00)	**-2.94** (0.45)	-0.90 **(5.41)**	0.26 (-1.33)	-0.55 **(-2.82)**	1.27 **(2.74)**	0.21 **(3.04)**	-1.05 (1.36)
Urugaryi (DJF)	-1.31 (0.69)	**-1.80** **(2.31)**	0.20 (0.00)	**-2.50** (0.38)	-1.17 **(5.19)**	1.49 (-0.63)	**-1.99** **(2.95)**	1.54 **(2.96)**	-0.12 **(3.25)**	0.52 (0.94)
Itumba (MAM)	-0.62 (1.03)	-0.15 **(3.11)**	-0.23 (0.15)	**-2.85** (0.98)	-0.71 **(4.26)**	0.79 (-1.19)	-1.06 **(-2.24)**	-0.95 **(2.08)**	0.56 **(2.70)**	0.00 (1.06)
Icyi (JJA)	-1.33 (1.54)	-0.26 **(2.90)**	0.48 (-0.19)	-1.12 (-0.38)	0.76 **(5.08)**	0.34 **(-2.48)**	0.21 **(-2.55)**	0.32 **(1.71)**	0.36 **(2.71)**	-0.84 (1.33)
Umuhindo (SON)	**-2.16** (0.69)	**-2.25** **(2.38)**	0.26 (0.00)	-0.49 (-0.61)	0.36 **(3.98)**	-0.50 (0.73)	1.53 **(-3.48)**	**2.37** **(1.70)**	-0.15 (2.86)	-1.22 (-0.09)

Table 3.5 demonstrates that recent mean temperature data exhibited a significant increasing trend for almost all stations. Most of the stations show increasing trends and somewhat statistically significant in mean temperature data at almost all data recorded recently (1971-2000 and 1961-2008).

An opposite behavior is observed at station Pt7 (Butare station) which is located in the case study of this research, Migina catchment (see Fig. 2.2). There was a statistically significant decreasing trend in all variables except in Urugaryi (DJF) seasonal mean temperature.

It is apparent from Table 3.5 that significant trends detected in the rainfall data tend to be fewer than in the temperature data. No significant trends were found in the majority of the stations for the analyzed rainfall data, and also greater differences between the trends are demonstrated. For example, there was a statistically significant decreasing trend in annual and DJF seasonal total rainfall at stations Pt2 and Pt4, while no trend was found in Icyi season (JJA).

Recent rainfall data (1971-2000 and 1961-2008) show a statistically significant decreasing trend from few stations. This was observed at station Pt1 in annual and Umuhindo (SON) seasonal total rainfall. Although, this decreasing trend in annual and Umuhindo (SON) seasonal rainfall was also observed at Pt2 include decreasing trends in Urugaryi (DJF) seasonal total rainfall as well. However, a decline is noticeable for data recorded recently (1971-2000) at Pt1 and Pt2 in annual rainfall as well as in all rainfall seasons with strong decline in short dry season Urugaryi (DJF) and short rain season Umuhindo (SON).

3.3.2. Streamflow trends and climate linkages

The relationship between climate and streamflow trends is shown in Figure 3.3 and Table 3.6. Kagera River at Rusumo station (St8) was selected for demonstration; it is located downstream of the Kagera study area.

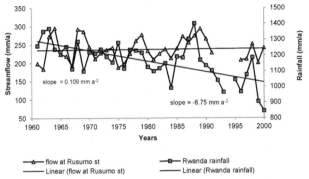

Figure 3.3 Inter-annual time series plots for areal rainfall and flow for Kagera and Nyabarongo Rivers. The triangle symbols show the observed streamflow values at Rusumo station (St8). The square symbols show the observed values for the Rwandan rainfall using Thiessen polygon method. Straight lines show the linear trend lines.

Figure 3.3 shows the data values for Rwandan average rainfall plotted against flow data for Kagera River at Rusumo station (St8). It is clear from Figure 4 that rainfall exhibits a decreasing trend (slope $= -6.75$ mm a^{-2}), while streamflow shows a slightly increasing trend (slope $= 0.11$ mm a^{-2}) at Rusumo station (St8) at the 95% confidence level. This means that stream flow series exhibits an opposite pattern at St8 over the period of records to that of Rwandan rainfall. The reason of this catchment behavior at St8 is probably due to the response of Ruvubu river basin to the Kagera river as explained in detail in Section 3.4.1. A clear response in rainfall-runoff can be seen at this station.

The positive relationship between rainfall and streamflow is observed and discussed in detail in Section 3.4.2. The correlation in streamflow in May at Rusumo station (St8) and average rainfall in March, April and May over all studied periods (1961-2000) are shown in Table 3.6.

Table 3.6 The correlation in streamflow in May at Rusumo station (St8) and average rainfall in March, April and May.

Time series	Av. Streamflow in May at St8 (m³ s⁻¹)	Av. Rainfall in March (mm month⁻¹)	Av. Rainfall in April (mm month⁻¹)	Av. Rainfall in May (mm month⁻¹)	Average rainfall in Mar-Apr-May (mm month⁻¹)	Correlation rainfall-runoff [-]
1961-2000	294.93	115.05	164.62	100.65	126.77	0.59

Table 3.6 shows the correlation between average flow in May and average rainfalls in March, May and April which is equal to around 60%. This shows that there is a linkage between May flow and both rainfalls in March (the calculated rainfall average is 115.1 mm month⁻¹) and April (the calculated rainfall average is 165.4 mm month⁻¹), which are previous months. The mean monthly rainfall in May is 100.7 mm for the whole studied periods (1961-2000). Similar results were found in almost all

studied stations and Kagera river is shown here due to its regional importance as it is located at downstream of the study area.

3.4. DISCUSSION OF RESULTS

3.4.1. Summary of trends

This section summarizes the noteworthy trends based on the two study periods (1971-2000 and 1961-2000), where four streamflow time series are tested for trends in 11 hydrological variables (Table 3.3). More significant trends are detected as the length of the study is increasing. This is probably due to the higher information content which is representing the dominant behavior of the catchments.

The observed increasing trends for St8 (Kagera catchment) might be attributed to the response (inflow) of the Ruvubu river (av. streamflow: 200 m^3 s^{-1}) from a large catchment (40%) inside Burundi (Kyaruzi et al., 2005), and contribute to the increase of Kagera river while rainfall is not appearing. For the shorter study periods (1971-2000), station St5 did not show significant trends in all selected variables. It can also be that there is just no trend, neither in rainfall nor in discharge observed.

The results for many of the monthly flow variables at all stations present increasing trends while results for many of the variables from annual monthly flow show both increasing and decreasing trends. We conclude that both increasing and decreasing trends of streamflow were detected in this study from 2 stations and 2 stations, respectively (Table 3.3). This confirms that an overall increasing trend of streamflow is detected in the longer period of data (1961-2000) and a decreasing trend in the short period of the study (1971-2000).

The observed shift upward of annual flows at St8 in 1962, while it is located downstream at St5 and St6 (see Fig. 2.2), should confirm to other inflow from Ruvubu river basin to the Kagera River as discussed below in this section. In general, the results of the Pettitt test reveal that most abrupt change points occurred in the period 1980s–1990s (Table 3.4). This period is related to a period of intensive human activities in Rwanda such as agriculture development (irrigation water use), urbanization, etc (PRSP, 2007). In addition, the observed changes in that period could also be due to the data quality because we cannot rely on data quality used in this study, which were collected in 1990 to 1994 due to the starting of war and genocide period in Rwanda. Probably data were not well observed in this period, or stations, which were destroyed, were not well rehabilitated due to an upward or downward shift and missing data were observed (see Tables 3.1, 3.2 and 3.4). The years of 1990 to 1995 present many gaps in the datasets and the IHA software interpolation of them. Maybe this linearly data filling is one of the reasons that most abrupt change points occurred in this period. Furthermore, this period was found to have unstable baseflow conditions; likely to be influenced by erosion/sedimentation processes of the river bed. Therefore, we have to doubt the quality of the data collected and used in this study, especially for the data collected in the highlighted period of war and genocide.

Rainfall totals over four different periods (1935-1993, 1960-1994, 1971-2000 and 1961-2008) at 10 selected stations was analyzed in this study. No significant trends were found in the majority of the stations at 5% significant level (Table 3.5).

Time series for minimum, mean and maximum temperature for the period from 1971 to 2006 at Kigali station were also analyzed (Table 3.5). Increasing trends were observed from almost all selected stations over the study period as demonstrated in Figure 3.2. The increasing temperature trends observed in Table 3.5 are possibly linked to human activities (e.g. deforestation, local urbanization) (MINITERE, 2006; REMA, 2009b; Munyaneza et al., 2011b; Safari, 2012), and can possibly attributed to natural variability or climate change (Safari, 2012).

Similar results were also found by RIWSP (2012b) in their study for the mean annual temperature at the same station Kigali (Pt5) during the past four decennia from 1971-2010. Mean annual temperature showed a significant trend of +0.47 °C/decade at the 10% significance level, which is close to our findings (+0.38°C/decade). Trends were not significant at the 5% significance level and we have detected some trends at the 10% significance level (Burn *et al.*, 2011). Unfortunately, temperature data for the period 2007-2010 were not available. This could be the reason for the slight differences found and gives a hint that, based on data analyzed from Pt1 to Pt10, the recent years have been getting warmer in Rwanda as also demonstrated by results given in Table 3.5 for different periods of analysis (e.g. Stations Pt2 and Pt5).

The observed warming was supported by Collins (2011) who demonstrated that significant increasing temperature trends were found on average all over the African continent. He used Microwave Sounding Unit (MSU) total lower-tropospheric temperature data from the Remote Sensing Systems (RSS) and the University of Alabama in Huntsville (UAH) data sets. The two most recent decades were compared with the period 1979-1990.

An opposite behavior (significant decreasing temperature trends) which was observed at station Pt7 (Butare station) could not be fully explained and more research should be done.

3.4.2. Relationship between hydrologic variables and climate variables

In this study, eight meso-scale catchments in Rwanda were investigated and during the rainy season period, the flows were generally increasing. The large flow observed in May flow is due to high rainfall of the previous months of March and April as a good correlation was observed ($r = 0.59$) (see Table 3.6).

Streamflow was found to be positively correlated with the total rainfall and negatively correlated with the mean temperature (Table 3.5), which is in line with the fact that temperature can be seen as a proxy for available energy in the system and, thus, for the atmospheric evaporation demand. The observed positive relationship between rainfall and streamflow is also supported by Mutabazi *et al.* (2004) in their study on the generation and application of climate information, products and services for disaster preparedness and sustainable development in Rwanda. However, the observed similarity in streamflow and rainfall variation implies that changes in stream flow should be linked with corresponding changes in the rainfall environment.

In the Kagera catchment, the monthly flows in March, April and May (MAM) are linked to the rainfall season (Table 3.6) due probably to the longer south-easterly monsoon as reported by Van Griensven *et al.* (2008) in their study on understanding riverine wetland-catchment processes in the Kagera river basin using remote sensing data and modelling. High stream flows in the investigated meso-scale catchments in Rwanda during the rainy season period were observed. For instance, the Nyabarongo river at St3 with an average streamflow of 122.3 m^3 s^{-1} in rainy season (MAM) while is 74.8 m^3 s^{-1} during dry season (JJA) for the studied periods (1961-2000). Highly positive correlation between rainfall and streamflows ($r = 0.58$, n = 40, p<0.05) was also observed during the rainy season (MAM) and no correlation ($r = -0.08$, n = 40, p<0.05) was observed in dry season (JJA).

3.5. CONCLUSIONS

The study identified the linkages between climatic data and streamflow data over selected meso-scale catchments in Rwanda (Sect. 3.4.2). The study shows differences for the different hydrologic variables and at specific locations and for the different study periods (Sect. 3.3.1). In general, trends in streamflow are stronger for the longer period of data (e.g. for 1961-2000 compared with 1971-2000). The results show significant increasing trends in temperature (1971 to 2006) and both increasing and decreasing trends (e.g. Pt2 and Pt4 in November) in rainfall (1961 to 2008) as demonstrated in Table 3.5.

In all stations, there are both increasing and decreasing trends and shift change points in many hydrological variables for the investigated catchments during the period 1980–1990 as shown in Table 3.4. The observed change points of most stations (Table 3.4) are likely due to human activities such as increased agricultural activities, irrigation, or water supply, but cannot be attributed to climatic changes (i.e. rainfall) (Zhang *et al.*, 2011). It could be also due to the fact that the observed correlation within the data was found to be statistically significant at a 5% confidence level for many stations. Streamflow was found to be positively correlated with the total rainfall and negatively correlated with the mean temperature due to the similar variation observed between them as also confirmed by Mutabazi *et al.* in 2004 (Table 3.5). However, the observed similarity in streamflow and rainfall variation implies that changes in streamflow should be linked with corresponding changes in the rainfall conditions. This gives a hint to the sensitivity of Rwanda water resources for future climatic conditions. This was supported by Pettitt tests, which reveal that abrupt change points of most stations occurred in recent 1980s–1990s (Table 3.4), which is related to the period of few data gaps (Table 3.1) and a period of intensive human development activities (PRSP, 2007). The existing gauging stations are providing unreliable data as rating curves were not updated (some of them are 30 years old and more). No regular discharge measurements are taken to update existing rating curves using Equation 2.1. Hydro-meteorological stations are not dense (even not well distributed) as it should be to capture all meteorological and hydrological data. Besides this, the results revealed that selected climate variables could not easily explain the observed trend behavior (see Table 3.5).

For future studies, we recommend a double-check on Rwandan data, especially on rating curves used to generate streamflows data (referring to Eq. 2.1). We have done our best for our results in this paper and we can stand behind our results. The selected 4 stations for data analysis had modified rating curves which were updated and corrected. Some modified rating curves were obtained for instance at St3 (RIWSP, 2012c), St5, St6 and St8 (RIWSP, 2012a). Therefore, we conclude that for stations St1, St2, St4, and St7 it is not possible to correct and adjust the raw data in such a way that further analysis regarding trend analysis could be carried out. New data with good quality are also needed for future better water resources management in Rwanda. Hence, the country needs much more to be sensitive about data collection.

Chapter 4

IDENTIFICATION OF RUNOFF GENERATION PROCESSES USING HYDROMETRIC AND TRACER METHODS

....

Understanding of dominant runoff generation processes in the meso-scale Migina catchment (257.4 km^2) in southern Rwanda was improved using analysis of hydrometric data and tracer methods. The paper examines the use of hydrochemical and isotope parameters for separating streamflow into different runoff components by investigating two flood events during the rainy season "Itumba" (March–May) over a period of 2 years at two gauging stations. Dissolved silica (SiO$_2$), electrical conductivity (EC), deuterium (^{2}H), oxygen-18 (^{18}O), major anions (Cl$^-$ and SO$^{2-}_4$) and major cations (Na$^+$, K$^+$, Mg^{2+} and Ca^{2+}) were analyzed during the events. ^{2}H, ^{18}O, Cl$^-$ and SiO$_2$ were finally selected to assess the different contributing sources using mass balance equations and end member mixing analysis for two- and three-component hydrograph separation models. The results obtained applying two-component hydrograph separations using dissolved silica and chloride as tracers are generally in line with the results of three-component separations using dissolved silica and deuterium. Subsurface runoff is dominating the total discharge during flood events. More than 80% of the discharge was generated by subsurface runoff for both events. This is supported by observations of shallow groundwater responses in the catchment (depth 0.2–2 m), which show fast infiltration of rainfall water during events. Consequently, shallow groundwater contributes to subsurface stormflow and baseflow generation. This dominance of subsurface contributions is also in line with the observed low runoff coefficient values (16.7 and 44.5%) for both events. Groundwater recharge during the wet seasons leads to a perennial river system. These results are essential for better water resources planning and management in the region that is characterized by very highly competing demands (domestic vs. agricultural vs. industrial uses).

....

Based on: Munyaneza, O., Wenninger, J., and Uhlenbrook, S., 2012. *Identification of runoff generation processes using hydrometric and tracer methods in a meso-scale catchment in Rwanda.* Hydrol. Earth Syst. Sci., 16: 1991–2004.

4.1 INTRODUCTION

Understanding the runoff components separation processes is essential for the proper assessment of water resources availability within catchments. The use of environmental isotopes in combination with hydrochemical tracers and hydrometric measurements can help to gain further insights into hydrological processes because the methods separate and quantify different runoff components during rainfall events. Combined methods can be used to quantify the contributions of runoff components during different hydrological situations (floods and low flows) in small and meso-scale catchments (Didszun and Uhlenbrook, 2008; Wenninger et al., 2008). Generally, hydrochemical and isotopic hydrograph separations of stream discharge are commonly used to determine the fractions of surface/subsurface or old/new water contributions to streamflow (e.g. Richey et al., 1998).

Most hydrograph separations involve the standard two-component mixing models of Sklash and Farvolden (1979), in which the stream water is separated into old (pre-event) and new (event) water components. This approach identifies the age of streamflow components, but cannot be used to assess the spatial origin (Ladouche et al., 2001). To obtain both temporal and spatial origins, some investigations using stable isotopes associated with chemical tracers, have been undertaken in different basins world-wide (for example, Kennedy et al., 1986; Wels et al., 1991; Ladouche et al., 2001; Uhlenbrook and Hoeg, 2003; Hrachowitz et al., 2011). However, hydrochemical tracers may only be used to separate streamflow into runoff components according to their flow paths (Kennedy et al., 1986).

Only few recent studies on the application of two and three-component hydrograph separation models improved our understanding of hydrological processes in semi-arid areas in Sub-Sahara Africa (Mul et al., 2008; Hrachowitz et al., 2011), where Rwanda is also located. These studies contribute to appropriately manage the available river water and groundwater resources, both in terms of quality and quantity. This is essential in Rwanda where the population is growing with an annual rate of about 3.5% (MINIPLAN, 2002), and it is already the most densely populated country on the African continent (NELSAP, 2007). The related increase of water demand for domestic, agricultural, and industrial uses is causing significant water scarcity in the country, and ecosystems are under enormous pressure.

Burns (2002) put it nicely by stating: "As the science matured further in the 1990s, a point was reached at which isotope-based hydrograph separations alone were insufficient to guarantee publication of study results in the leading water resources journals. Many studies seemed only to reconfirm that stormflow in small forested catchments is dominated by 'pre-event' or 'old' water, and hydrologists did not need to be told so over and over again. Thus, isotope-based hydrograph separation had become simply another tool - one that could not lead to a more profound understanding of catchment runoff processes unless combined with many other tools." Since then, the application of hydrograph separation together with hydrometric observation became state of the art in the global North, but much less in the global South in particular in remote area of Africa with its unique hydro-climatic and other physiographic settings. However, hydrograph separation methods were applied before to semi-arid or better sub-humid catchments with the support of well data (Cras et al., 2007; Marc et al., 2011; Hrachowitz et al. 2011), but these studies site are different than the study area in Rwanda.

Detailed insights into the hydrology of a meso-scale catchment like the Migina catchment contributes to an increased understanding regarding the water resources of the catchment; an important first step is ensuring a sustainable level of development in the future. However, the role of catchments in understanding hydrologic processes can be explained with better agriculture and forest management as well as other development activities conducted within catchments (Peters, 1994). This knowledge can help famers to increase their crop production and to sustain long-term food security (e.g. Mul, 2009; Hrachowitz et al., 2011). In order to achieve this, insights into the behavior of the water fluxes and the interactions between groundwater and river water is of utmost importance. Munyaneza et al. (2011b) conducted their study in the meso-scale Migina catchment, southern Rwanda, to predict river flows. Van den Berg and Bolt (2010) also conducted their research in the same catchment using hydrochemical and isotope analysis during the dry season (June to August 2009). Based on a baseflow recession curve analysis, they showed a decreasing trend in baseflow in the overall river discharge. It is now becoming almost constant at a rate of 0.19 m^3 s^{-1} at the main outlet at the end of the dry season. Furthermore, they concluded that a significant flow from (deep) groundwater has to be the source of this water. Hence, the suggestion was made to perform detailed hydrochemical and isotopic hydrograph investigations also during floods to obtain a better understanding of groundwater-surface water interactions as well as the different sources and flow pathways. Burns (2002) found that the thrill of doing isotope-based hydrograph separations in forested, humid catchments is gone. Therefore, he recommended carrying out new studies in catchments with different climatic and human disturbance regimes. Additionally, these studies which combine water-isotope and solute isotope measurements should provide hydrologists with new thrills and even surprises in the coming years. Consequently, the current study was carried out in a humid catchment and contributed to the advancement of hydrologic science of this hydro-climatic zone by quantifying runoff components and processes. Hardly any studies can be found in related hydro-climatic zones in the literature; therefore, we feel this study is a good addition to the existing knowledge base.

The objective of the paper is to quantify the runoff components and to identify the dominant hydrological processes in a meso-scale catchment for two flood events occurred during the rainy season "Itumba" (March–May) over a period of 2 years, i.e. 1 to 2 May 2010 at the outlet of the Cyihene-Kansi sub-catchment and 29 April to 6 May 2011 at the outlet of the Migina catchment in southern Rwanda (Fig. 2.4). Specifically, the study emphasizes the use of two- and three-component hydrograph separation mixing models for separating streamflow into surface and subsurface runoff and quantifying different runoff components under tropical conditions. In order to learn more about hydrologic flow paths, hydrochemical tracers and hydrometric measurements such as rainfall, stream discharge, springs and groundwater levels were combined with tracer studies. The study explores the importance of combining hydrometric data, isotope information and hydrochemical tracers to identify runoff components (e.g. Ladouche et al., 2001; Uhlenbrook et al., 2002).

4.2 DATA AND METHODS

4.2.1 Data collection

The catchment has been equipped with hydrological instruments (Fig. 2.4) and after installation, hydrochemical and isotope data were collected over two years (May 2009 to June 2011). Two events were examined in further detail during the long rainy season "Itumba". Intensive monitoring (hourly and daily samples) was carried out between 1 and 2 May 2010 and between 29 April 2011 and 6 May 2011 at Kansi and Migina gauging stations, respectively. Samples were analyzed in the lab for isotopes and hydrochemical tracers. The collected samples include groundwater from 11 shallow piezometers, 15 springs, river discharge measurements from 5 river gauging stations (Rwabuye, Mukura, Kansi, Akagera, and Migina); stream water sampled at 8 sites in the catchment (weekly or monthly intervals), and monthly catchment rainfall from 5 locations where tipping buckets are installed (see Fig. 2.4). One rainfall event during the Itumba'11 season (from 29 April 2011 to 6 May 2011) was also sampled at Gisunzu rain gauge for isotopic composition analysis.

4.2.2 Field and laboratory methods

In-situ measurements have been continuously conducted at the outlet of each sub-catchment for pH value and water temperature (T) using a portable pH-meter (Hach 157) and for electrical conductivity (EC) using a Hanna Gro'Chek Portable EC-meter (HI9813-0). Stream, spring and rain water samples were collected in 30 ml plastic bottles. Samples were collected during low flows and flood events.

Samples were analyzed in the laboratory for dissolved silica (SiO_2) using a Spectrophotometer DR 2400 at the laboratory of Kadahokwa water treatment plant and at the laboratory of the University of Rwanda (UR), Butare, Rwanda. The concentrations of major cations like Mg^{2+}, Ca^{2+} and K^+ were determined by Atomic Absorption Spectroscopy (AAS) at UR-Huye Campus and sodium (Na^+) was determined by AAS at UNESCO-IHE, Delft, The Netherlands. The concentrations of major anions such as SO^{2-}_4 were determined using a Hach-DR/890 Colorimeter in the laboratory of WREM at UR, and Cl^- was analyzed by using an Ion Chromatograph at UNESCO-IHE and verified by using Colorimetry in the laboratory of UR-Huye Campus. The isotopes were analyzed at UNESCO-IHE with a LGR Liquid-Water Isotope Analyzer, which provides measurements of $\delta^{18}O$ and δ^2H in liquid-water samples with accuracy better than 0.2‰ for $^{18}O/^{16}O$ and better than 0.6‰ for $^2H/^1H$.

During the investigated two flood events, the water levels were measured continuously at two river gauging stations (Kansi and Migina) using pressure transducers (Mini-Diver; DI501) and transferred to discharges using rating curves ($r^2 = 0.94$, $n = 24$ at Kansi station and $r^2 = 0.97$, $n = 18$ at Migina station). The above rating curves were calculated using Equation (2.1).

4.2.3 Hydrometric and tracer methods

Hydrograph separation to separate the runoff during floods in two or more components (end-members), based on the mass balances for tracer fluxes and water, was applied in this study. Environmental isotopes (oxygen-18 (^{18}O) and deuterium (^2H)), dissolved silica (SiO_2) and chloride (Cl^-) were selected as tracers.

The fundamentals and assumptions of the hydrograph separation method are further discussed in e.g. Sklash and Farvolden (1979), Wels et al. (1991), Buttle (1994) and Uhlenbrook and Hoeg (2003). The mass balance expression for a two-component hydrograph separation model used in this paper is described as follows:

$$Q_T = Q_1 + Q_2 \tag{4.1}$$

$$c_T Q_T = c_1 Q_1 + c_2 Q_2 \tag{4.2}$$

where: Q_T is the total runoff ($m^3\ s^{-1}$); Q_1, Q_2 are runoff contributions ($m^3\ s^{-1}$); c_T is the concentration in the total ($mg\ l^{-1}$ or ‰); and c_1, c_2 are the end-member concentrations of the tracers in the respective runoff component ($mg\ l^{-1}$) or (‰).

The exact definition of the two or three runoff components depends on the properties of the tracer used (Wels et al., 1991). Two commonly used groups of tracers are: (1) stable isotopes of water, oxygen-18 (^{18}O) and deuterium (^2H) (e.g. Sklash and Farvolden, 1979; Sklash et al., 1986) and (2) weathering products such as Mg^{2+}, Ca^{2+}, Cl^- and SiO_2 (e.g. Pinder and Jones, 1969; Wels et al., 1991).

With a known concentration of the end-members for subsurface and surface runoff, the contribution from these sources can be calculated (e.g. Mul et al., 2008). The concentration for sub-surface (including groundwater) runoff was assumed to be the concentration of the pre-event water at the sampling point and the concentration of the surface runoff was assumed to be similar to concentrations observed in a rainfall sample (Buttle, 1994; Mul et al., 2008). These end member concentrations were based on the observed dilution of river runoff during rainfall events. Therefore, the total discharge Q_T and concentrations c_T, c_1 and c_2 are known and it follows:

$$Q_2 = \frac{c_T - c_1}{c_2 - c_1} Q_T \tag{4.3}$$

$$Q_1 = Q_T - Q_2 \tag{4.4}$$

Hrachowitz *et al.* (2011) applied hydrochemical tracers in combination with isotopic tracers for hydrograph separation in a semi-arid catchment in Tanzania. They found that the assumption of stable isotopic end-members was not met for both the groundwater samples and the rain water samples. At the small scale the spatial variability could be negligible and the technique becomes better applicable, although for each event, end-member concentrations needed to be determined separately to account for the temporal variability. Due to this temporal variation, hydrograph separation was performed in this paper using the cumulative incremental weighting approach, Eq. (4.5), based on sampled rainfall amount as recommended by McDonnell *et al.* (1990):

$$\delta^{18}O = \frac{\sum_{i=1}^{n} P_i \delta_i}{\sum_{i=1}^{n} P_i} \tag{4.5}$$

where: P_i and δ_i denote fractionally collected rainfall amounts and δ values (isotope concentrations), respectively. The weighted mean represents the average isotopic composition of the event water input to the catchment but does not address the within-storm isotopic variability or the time response of the catchment to event water (McDonnell *et al.*, 1990).

A three-component hydrograph separation was applied in this study by using dissolved silica and deuterium for the event of 1–2 May 2010 at Kansi station (Fig. 4.6) and using dissolved silica and oxygen-18 as tracers for the event of 29 April 2011 to 6 May 2011 at Migina station (Fig. 4.8). The same method was used by James and Roulet (2009) to estimate the relative contributions of throughfall, a perched groundwater or shallow subsurface flow component and groundwater for individual storm events in small forest catchments of Mont Saint-Hilaire in Quebec, Canada. During our research, three end-members (pre-event: deep and shallow groundwater, and event: rainfall) were used in the separation. End-member concentrations were collected for each event separately in order to account for the temporal variability as accurately as possible. The end-member for deep groundwater was selected to be the one from springs and from deep piezometers installed in hillslope. Shallow piezometers close to stream were considered to represent the end member of shallow groundwater. The end-member concentration for rainfall was taken as average rainwater sampled at four automatic (tipping buckets) rainfall stations installed in the study area (see Fig. 2.4).

Event-based runoff coefficient estimations were determined from Thiessen polygon representation of rainfall and continuous runoff records (Burch *et al.*, 1987; Iroumé *et al.*, 2005; Blume *et al.* 2007). In the study presented here, the runoff coefficient for each event was computed by dividing the total flow by the total rainfall as recommended by Spieksma (1999) and Iroumé *et al.* (2005). Using total flow allows us to combine the response of the single event with the pre-event flow conditions (Blume *et al.*, 2007). Rainfall measurements have been carried out by using 13 manual rain gauges installed in the Migina catchment. The endpoint of each event has been estimated by waiting until the discharge is back to baseflow conditions. This did not cause very long tailings (recession limbs) for the event probably due to the short catchment response of 3.5 hours observed by Munyaneza *et al.* (2011b) in the same catchment.

4.3 RESULTS
4.3.1 Rainfall-runoff observations for Itumba'10 & 11 seasons (March-May)

The observed discharges in the center of the Migina catchment at Kansi station, for data recorded from 1 May 2009 to 31 June 2011, were in the range of 0.24–9.16 m³ s⁻¹ and average discharge was estimated to 1.71 m³ s⁻¹. The observed discharges at the outlet of the Migina catchment (at Migina station), for data recorded from 1 August 2009 to 31 June 2011, were in the range of 0.43–15.60 m³ s⁻¹ with an average discharge of 3.35 m³ s⁻¹.

Rainfall measurements have been done at 13 manual rain gauges installed in the Migina catchment, i.e. the Gisunzu and Murama rain stations were not considered for the areal rainfall of the Cyihene-Kansi sub-catchment (see Fig. 2.4). The amount of rainfall in both Cyihene-Kansi and Migina catchments were estimated using the Thiessen polygons method, which seems appropriate due to spatial distribution of the rainfall stations and the low topographic gradients.

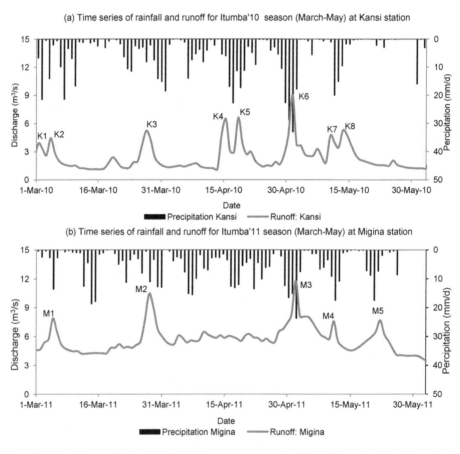

Figure 4.1 Time series of rainfall and runoff events during March-May 2010 at Kansi station (a) and March-May 2011 at Migina station (b).

Figure 4.1 shows the rainfall and discharge patterns observed at Kansi (Fig. 4.1a) and Migina (Fig. 4.1b) gauging stations during the investigated periods (Itumba'10 and Itumba'11). The detailed investigated two flood events are event K6 for Kansi station and event M3 for Migina station (Tables

4.1 and 4.2). Seasonal rainfall totals to 552 mm and 508 mm for Cyihene-Kansi sub-catchment and Migina catchment, respectively. These seasonal rainfall totals generate on average a runoff of 2.42 m^3 s^{-1} (148.7 mm) at Kansi station, and 5.75 m^3 s^{-1} (177.7 mm) at Migina station.

The time series of rainfall and runoff for storm event K6 and M3 represent the intensive monitoring periods in this research. Maximum daily rainfall of 32.9 mm d^{-1} (4.3 x 10^6 m^3) was observed on 2 May 2010 at the outlet of the Cyihene-Kansi sub-catchment and the runoff generated by this rainfall at Kansi station reaches its peak at the same day at 03:00 LT (9.05 m^3 s^{-1}). Peak flows are estimated based on instantaneous river runoff observations recorded using Mini-Diver (DI501) in 15 min interval (see Tables 4.1 and 4.2). The river discharge returns to pre-event values on 5 May 2010 when the surface runoff contribution stopped. Similarly, a maximum daily rainfall of 23.7 mm d^{-1} (6.1 x 10^6 m^3) was observed on 2 May 2011 at the outlet of the Migina catchment and the runoff generated by this rainfall at Migina station, reaches its peak at 10:00 LT (11.78 m^3 s^{-1}) the same day. The river discharge returns to pre-event values on 6 May 2011.

Table 4.1 Rainfall-runoff events during Itumba'10 season in the Cyihene-Kansi sub-catchment (129.3 km^2). The detailed investigated event K6 is given in bold.

Event N°	Date	Time	Rainfall event		Runoff event					
			Duration	Maximum rainfall intensity	Rainfall amount	Peak runoff	Specific peak runoff	Runoff volume	Total Runoff	Runoff coef.
			(h)	(mm h^{-1})	(mm)	(m^3 s^{-1})	(mm h^{-1})	(10^4 m^3)	(mm)	(%)
K1	2-Mar	7:05	8.0	2.0	41.98	3.91	0.109	119.5	9.24	22.0
K2	5-Mar	4:20	7.0	0.8	27.92	4.47	0.124	144.0	11.13	39.9
K3	28-Mar	10:35	7.0	5.6	70.09	5.23	0.146	229.9	17.78	25.4
K4	16-Apr	7:35	8.0	11.2	74.04	6.47	0.180	159.9	12.37	16.7
K5	19-Apr	10:50	11.3	9.2	79.51	6.63	0.185	293.5	22.70	28.5
K6	**2-May**	**3:00**	**22.0**	**16.6**	**113.27**	**9.05**	**0.252**	**265.0**	**20.49**	**18.1**
K7	11-May	23:50	5.5	10.6	47.12	4.69	0.131	120.6	9.32	19.9
K8	14-May	18:20	6.0	3.6	50.57	5.26	0.147	291.3	22.53	44.5

Table 4.2 Rainfall-runoff events during Itumba'11 season in the Migina catchment (257.4 km^2). The detailed investigated event M3 is given in bold.

Event N°	Date	Time	Rainfall event		Runoff event					
			Duration	Maximum rainfall intensity	Rainfall amount	Peak runoff	Specific peak runoff	Runoff volume	Total Runoff	Runoff coef.
			(h)	(mm h^{-1})	(mm)	(m^3 s^{-1})	(mm h^{-1})	(10^4 m^3)	(mm)	(%)
M1	5-Mar	9:38	11.0	12.0	75.87	7.89	0.110	615.8	23.92	31.5
M2	28-Mar	0:08	6.2	14.8	49.87	10.46	0.146	570.5	22.16	44.4
M3	**2-May**	**10:00**	**14.0**	**17.6**	**96.32**	**11.78**	**0.165**	**883.6**	**34.32**	**35.6**
M4	11-May	3:51	2.5	7.6	42.47	7.57	0.106	421.4	16.37	38.5
M5	22-May	2:20	10.0	9.4	54.31	7.69	0.108	447.3	17.37	32.0

Tables 4.1 and 4.2 show the main hydrological characteristics of 8 different events during Itumba'10 and 5 different events monitored during Itumba'11 at Kansi and Migina gauging stations, respectively. Runoff coefficients were observed ranging from 16.7% to 44.5% with maximum rainfall intensities up to 16.6 mm h^{-1} for Itumba'10 and 17.6 mm h^{-1} for Itumba'11.

Most rain events during both seasons Itumba'10 and Itumba'11 are moderate (2.5 to 7.5 mm h^{-1}) or heavy (>7.5 mm h^{-1}). Only light rain was observed on 2 March 2010 at 07:05 (2.0 mm h^{-1}) and on 5 March 2010 at 04:20 (0.8 mm h^{-1}) for the Itumba'10 season (Table 4.1). The observed low runoff coefficients, for Cyihene-Kansi sub-catchment (16.7–44.5%) and Migina catchment (31.5–44.4%) indicate that a high percentage of the rainfall becomes subsurface runoff. This is later supported by the hydrograph separation (see Sect. 4.4.2). Rainfall amount and runoff volume show a strong correlation ($r = 0.93$, $n = 18$) for Cyihene-Kansi sub-catchment and ($r = 0.95$, $n = 19$) for Migina catchment.

4.3.2 Results of hydrochemical tracer studies

The most important hydro-chemical parameters of the water samples from springs, rivers, rainfall and shallow groundwater wells are presented in Table 4.3.

Table 4.3 Hydrochemical concentrations observed in the Cyihene-Kansi sub-catchment and Migina catchment during the investigated research period (from 1 May 2009 to 31 June 2011). n represents the number of samples. The entries in brackets represent the range values.

	Parameter	Unit	Rainfall (n = 103)		River water (n = 173)		Groundwater (n = 59)		Springs (n = 34)	
			Kansi	Migina	Kansi	Migina	Kansi	Migina	Kansi	Migina
	pH	-	6.0 (5.7-6.8)	6.1 (5.3-6.9)	6.9 (5.8-7.3)	6.8 (5.6-7.8)	6.0 (5.1-9.4)	6.0 (5.1-9.4)	5.0 (3.7-6.2)	5.1 (4.1-6.3)
	EC	$\mu S\ cm^{-1}$	67.7 (54.2-227.4)	52.3 (42.3-271.4)	99.1 (82.6-305.3)	135.5 (73.8-452.9)	217.3 (146.6-450.0)	217.3 (146.6-450.0)	131.7 (56.3-143.5)	127.6 (52.3-158.0)
	SiO_2	$mg\ l^{-1}$	2.8 (0.0-15.5)	1.8 (0.0-9.3)	8.8 (0.0-35.0)	11.3 (0.0-19.7)	16.2 (9.0-31.0)	16.2 (9.0-31.0)	21.7 (16.2-28.6)	22.9 (17.3-37.7)
Anions	SO_4^{2-}	$mg\ l^{-1}$	1.2 (0.0-7.0)	1.3 (0.0-7.0)	8.3 (3.0-12.0)	8.4 (4.6-12.2)	9.2 (1.1-54.1)	9.2 (1.1-54.1)	3.1 (2.2-9.2)	5.0 (2.2-7.9)
	Cl^-	$mg\ l^{-1}$	0.52 (0.3-5.9)	1.0 (0.5-3.6)	4.16 (2.1-13.8)	6.4 (2.7-12.9)	1.2 (0.0-5.0)	1.2 (0.0-5.0)	5.6 (0.4-9.9)	5.6 (2.1-11.7)
Cations	K^+	$mg\ l^{-1}$	1.0 (0.4-4.3)	1.5 (0.3-5.2)	1.1 (0.6-2.2)	1.3 (1.2-1.7)	3.3 (0.3-5.6)	3.3 (0.3-5.6)	2.1 (1.7-5.1)	3.2 (2.2-5.4)
	Mg^{2+}	$mg\ l^{-1}$	0.3 (0.02-1.3)	0.5 (0.02-2.7)	1.9 (0.7-2.9)	2.5 (2.1-3.1)	2.9 (1.1-4.3)	2.9 (1.1-4.3)	3.2 (1.7-4.4)	3.4 (2.7-5.1)
	Ca^{2+}	$mg\ l^{-1}$	0.7 (0.1-3.6)	1.5 (0.19-3.1)	3.2 (1.5-5.6)	5.0 (2.9-5.3)	13.7 (4.5-17.4)	13.7 (4.5-17.4)	10.1 (4.0-12.8)	8.8 (4.4-12.0)
	Na^+	$mg\ l^{-1}$	-	24.4 (10.0-28.9)	-	36.4 (9.1-44.2)	55.7 (9.2-74.0)	55.7 (9.2-74.0)	6.7 (3.9-9.4)	6.1 (6.1-6.6)

Table 4.3 shows that the concentrations of most of the chemical components in river water are related to the concentrations of water sampled from springs and piezometers during flood events. The opposite can be seen in dissolved silica (SiO_2) and electrical conductivity (EC) concentrations decrease during peak flow. This indicates that river discharge is dominated by subsurface runoff components during flood events in the Migina catchment. This agrees with the low runoff coefficients observed in the catchments (Tables 4.1 and 4.2). Range (minimum and maximum) values in Table 4.3 were calculated based on measurements.

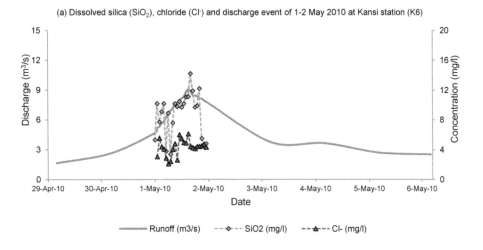

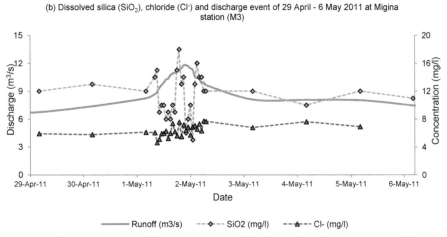

Figure 4.2 Hydrochemical parameter responses at Kansi station from 12:00 of 1 May 2010 to 11:00 of 2 May 2010 storm event (a) and at Migina station from 16:00 of 29 April 2011 to 16:00 of 6 May 2011 storm event (b).

Figure 4.2 shows the concentrations of dissolved silica and chloride during the two investigated events. The hydrograph is rising from 2.6 m^3 s^{-1} to 9.1 m^3 s^{-1} at Kansi river and from 6.5 m^3 s^{-1} to 11.8 m^3 s^{-1} at the outlet of the Migina catchment. Unfortunately, baseflow was not sampled for the season Itumba'10 (Fig. 4.2a) but sampled for season Itumba'11 (Fig. 4.2b).

Hourly SiO$_2$ and Cl$^-$ concentrations observed in stream water during the event of 1 to 2 May 2010 do not show clear trends but a small increase was observed during the peak flow and followed by constant concentrations for Cl$^-$, and smooth recession towards background concentration for SiO$_2$ (Fig. 4.2a). The observed concentrations during low flows for season Itumba'11 do not present clear trends either but increase and decrease near the peak can be seen during the flood event (Fig. 4.2b). This means that the hydrochemical parameters (SiO$_2$ and Cl$^-$) show a similar behavior for this event and remain constant during low flows, between 10–12 mg l^{-1} for SiO$_2$ and 5.8–7.6 mg l^{-1} for Cl$^-$. Distinct variations were observed during flood events, between 4–18 mg l^{-1} for SiO$_2$ and 4.6–7.7 mg l^{-1} for Cl$^-$ (Fig. 4.2b).

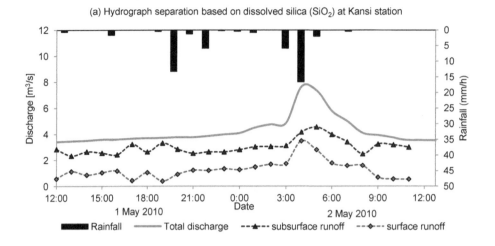

(a) Hydrograph separation based on dissolved silica (SiO$_2$) at Kansi station

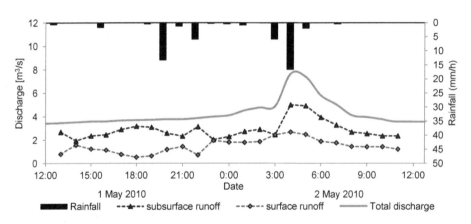

(b) Hydrograph separation based on chloride (Cl$^-$) at Kansi station (K6)

Figure 4.3 Results of two-component hydrograph separations based on dissolved silica (a) and chloride (b) for subsurface and surface runoff for event K6 (see Fig. 4.1a) investigated from 12:00 of 1 May 2010 to 11:00 of 2 May 2010 at Kansi station.

Figure 4.3 demonstrates that hydrograph separations using dissolved silica (Fig. 4.3a) and chloride (Fig. 4.3b) as tracers show that subsurface runoff during the event on 2 May 2010 is dominating the surface runoff and contributes from 54 to 89% (about 75% on average) and from 50 to 85% (about 70% on average), respectively. This confirms the observation of low contribution of direct surface runoff, supported by low runoff coefficients (Tables 4.1 and 4.2). Due to the fact that the whole rising limb, peak and recession limb were not captured completely for this event, the entire streamflow generated by groundwater could not be quantified. However, the dominance of subsurface runoff was observed during the starting time of the event sampling and subsurface runoff contributed 77.2%, which allows concluding that the overall contribution of surface runoff is relative small. The fact that surface runoff could be detected even before the main event is due to rainfall distribution during the rainy season that triggered some localized surface runoff generation and (delayed) inflow to the river throughout the season.

The observed maximum contributions of surface runoff during the peak flows are not equal in terms of timing for the separations using dissolved silica (SiO_2) and chloride (Cl^-). Using SiO_2 the maximum surface runoff contribution (45%) was observed on 2 May 2010 at 15:00 LT, then one hour later the peak runoff was reached at 16:00 LT while using Cl^- about 50% of this contribution was observed at the same time as the peak runoff (on 2 May 2010 at 15:00 LT). This timing difference can be attributed to various uncertainties related to the method (cf. methods section) and should not be over-interpreted. The observed subsurface runoff dominance is also supported by the findings of Munyaneza *et al.* (2011b) who showed that groundwater in the Migina catchment is very shallow (depth between 0.2–2 m in the valleys) and infiltrated rain water can reach the groundwater quickly and contribute to subsurface stormflow and baseflow during and after events, respectively. The depth can reach up to 4.1 m at the hilltops as found by Van den Berg and Bolt (2010).

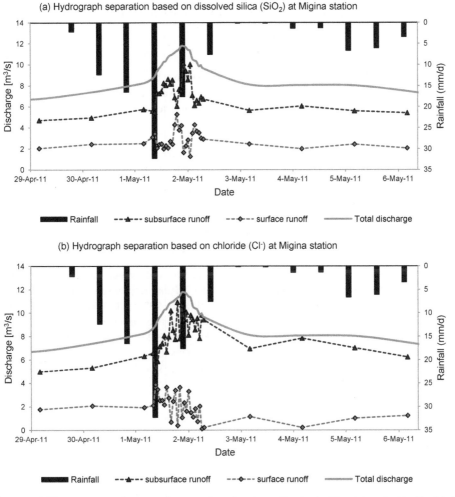

Figure 4.4 Two-component hydrograph separations based on dissolved silica (a) and chloride (b) for subsurface and surface runoff for event M3 (see Fig. 4.1b) investigated from 16:00 of 29 April 2011 to 16:00 of 6 May 2011 at Migina station.

Figure 4.4 shows the hydrograph separations using dissolved silica (Fig. 4.4a) and chloride (Fig. 4.4b) as tracers during the event of 29 April 2011 at 16:00 to 6 May 2011 at 16:00 at Migina station. The results are similar as the separations for event of 1 May 2010 at 12:00 to 2 May 2010 at 11:00 at Kansi station. Subsurface runoff is dominating the surface runoff and contributes from 53 to 89% (about 75% on average) and from 56 to 99% (about 80% on average) using dissolved silica and chloride, respectively.

The results of the two-component hydrograph separations show that the majority of the flood was generated by subsurface runoff (80%) and the surface runoff contribution hardly varies during the event except some increase during the peak time. Similar to the event of May 2010 (Fig. 4.3), the maximum contribution of surface runoff during the event of May 2011 was observed at slightly different times for both tracers. Using dissolved silica for hydrograph separation, maximum surface runoff contribution was observed three hours before the peak runoff was reached (on 2 May 2011 at 07:00 LT) and contribute 47%, while for chloride the maximum was observed two hours before the peak runoff was reached (on 2 May 2011 at 08:00 LT) and contribute up to 44%. The falling limb is largely dominated by subsurface runoff.

4.3.3 Results of isotopes tracer studies

The assumptions of hydrograph separation (Sect. 4.2.3) have been investigated by comparing the temporal and spatial variability of the different tracers in rain water and groundwater from springs and piezometers. In other words, the stability of end members was tested for the application of the three-component hydrograph separation technique.

Table 4.4 Isotope concentrations observed at the Cyihene-Kansi sub-catchment and at the Migina catchment during the investigated research period (from 1 May 2009 to 31 June 2011); n represents the number of samples; the entries in brackets represent the range values.

	Parameter	Unit	Rainfall (n = 145)		River water (n = 173)		Groundwater (n = 28)		Springs (n = 18)	
			Kansi	Migina	Kansi	Migina	Kansi	Migina	Kansi	Migina
Isotopes	$\delta^2 H$	(‰)	-16.9 (-55.2-13.8)	-7.8 (-42.5-(19.2))	-11.4 (-25.6-(-2.4))	-3.5 (-16.5-(-0.6))	-15.2 (-20.0-(-11.2))	-15.2 (-20.0-(-11.2))	-9.4 (-10.5-(-9.0))	-8.8 (-8.2-(-5.8))
	$\delta^{18} O$	(‰)	-4.3 (-8.9-(-0.4))	-3.3 (-7.5-(0.35))	-3.0 (-4.9-(-1.2))	-1.5 (-3.4-(-1.1))	-3.7 (-4.3-(-3.1))	-3.7 (-4.3-(-3.1))	-3.1 (-3.6-(-3.0))	-3.2 (-3.3-(-3.0))

Table 4.4 shows that the mean values of $\delta^2 H$ and $\delta^{18} O$ in river water runoff are −11.4‰ and −3.5‰ for $\delta^2 H$; and −3.0‰ and −1.5‰ for $\delta^{18} O$ at Kansi and Migina, respectively. The values of these isotopes in rainfall water are −16.9‰ and −7.8‰ for $\delta^2 H$; and −4.3‰ and −3.3‰ for $\delta^{18} O$. The mean values of $\delta^2 H$ and $\delta^{18} O$ were also investigated in the same two catchments (Cyihene-Kansi and Migina) during the entire period of research (May 2009–June 2011) for groundwater during floods and low flows. Their values in shallow groundwater obtained from piezometers are −15.2‰ and −3.7‰, respectively. The mean values of $\delta^2 H$ and $\delta^{18} O$ in water sampled from springs are −9.4‰ and −8.8‰ and −3.1‰ and −3.2‰, respectively.

End-member concentrations for deep and shallow groundwater were estimated based on data from piezometers located in the upper part of a hillslope and in a near stream location (Munyaneza *et al.*, 2010). The end-member for rainfall samples was taken as an average of rainwater sampled at 4 automatic rainfall stations (see Fig. 2.4).

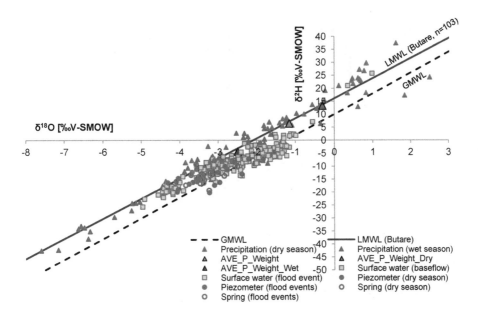

Figure 4.5 Stable isotope compositions of rainfall, river water, springs, shallow groundwater, and amount weighted rainfall for dry and wet seasons. GMWL: $\delta^2H = 8.13\ \delta^{18}O + 10.8$ (Source: Clark and Fritz, 1997). GMWL is the Global Meteoric Water Line; LMWL is the Local Meteoric Water Line for Butare; AVE_P_Weight means the average weight rainfall concentration for water sampled during wet and dry seasons; AVE_P_Weight_Dry means the average weight rainfall concentration for water sampled in summer season; and AVE_P_Weight_Wet represents the average weight rainfall concentration for water sampled during in rainy season.

Figure 4.5 shows stable isotopes (oxygen-18 (^{18}O) and deuterium (^2H)) in the water sampled in the Cyihene-Kansi sub-catchment and Migina catchment during the 2-year study period. The slope of the constructed Local Meteoric Water Line for Butare (LMWL Butare, $\delta^2H = 7.72 \cdot \delta^{18}O + 16.12‰$; $n = 103$) is close to the one of the Global Meteoric Water Line (GMWL, $\delta^2H = 8.13 \cdot \delta^{18}O + 10.8‰$), but has a clearly higher intercept. However, it is obvious in Figure 4.5 that the wet season rainfall is responsible for the light values of the groundwater and the baseflow. The isotopic composition of the rainfall is clearly different in the dry and wet season, and the wet season rainfall signature dominates the other water balance components (surface and subsurface water). Interestingly, the isotope values of the observed springs are not influenced by dry season rainfall values, as they all plot below the LMWL, show lighter isotope values than the amount weighted rainfall values of the wet season rainfall input. Thus, it can be concluded that the perennial springs in the area are recharged exclusively during the wet season.

The figure shows also that most of the stables isotopes of groundwater and spring water in the catchments are lighter than those of the stream waters and they plot even below the LMWL. This means probably that infiltrated water is affected by evaporation before reaching the groundwater system (temporary storage in soil zone). Similar results were found for instance by Kabeya *et al.* (2007) in a forested watershed in Kampong Thom, Cambodia.

A three-component hydrograph separation was applied in this study by using dissolved silica and deuterium for the event of 1–2 May 2010 at Kansi station (Fig. 4.6) and using dissolved silica and oxygen-18 as tracers for the event of 29 April 2011 to 6 May 2011 at Migina station (Fig. 4.8).

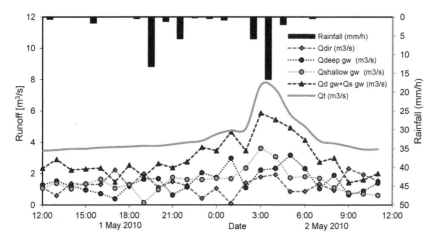

Figure 4.6 Results of the three-component separation using dissolved silica and deuterium as tracers for event K6 (see Fig. 2a) investigated from 1 May 2010 at 12:00 to 2 May 2010 at 11:00 at Kansi station. $Q_{dgw} + Q_{sgw}$ is the sum of deep and shallow groundwater components.

Figure 4.6 shows the results of the three-component separation method using dissolved silica and deuterium as tracers for the investigated event of 2 May 2010 at Kansi station. The results are comparable to the results obtained from the two-component hydrograph separations (see Sect 4.3.2). Pre-event water (deep and shallow groundwater, $Q_{dgw} + Q_{sgw}$) is dominating the discharge generation in this event and is contributing 38–98% (about 80% on average) to the total discharge (Q_t). Event water (direct runoff, Q_{dir}) dominates during few hours (on 1 May 2010 at 17:00 LT) during the rising limb and contributes then about 60%. The peak flow is also dominated by pre-event water (76.7%) and occurred on 2 May 2010 at 03:00 LT. Note that the shallow groundwater has been sampled in the valley, and the deep groundwater has been observed at perennial springs with constant discharge and hydrochemical characteristics.

The rainfall was sampled intensively during the event of 29 April 2011 to 6 May 2011 with a high temporal resolution of rainfall samples for isotope analysis (Fig. 4.7). The $\delta^{18}O$ value of the rainfall event ranges between −1.93‰ to −1.24‰ and the mean bulk rainfall $\delta^{18}O$ value for the whole event is equal to −1.52‰ (see Fig. 4.7). The incremental weighting approach based on rainfall amount was applied, Eq. (4.5), as recommended by McDonnell *et al.* (1990), but due to the observed very low temporal variations of isotopes in rainfall, the effect of this method is limited.

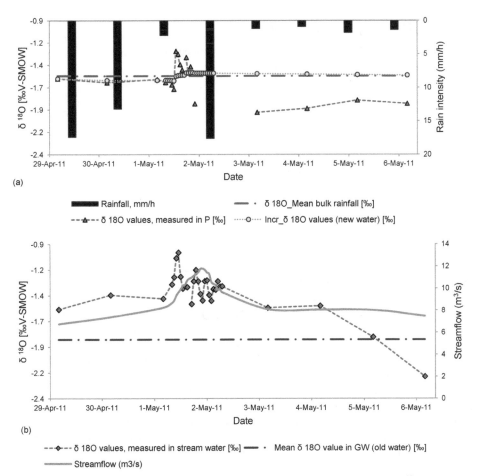

(a)

(b)

Figure 4.7 Hourly rainfall and variations of $\delta^{18}O$ in rainfall (a), discharge and variations of $\delta^{18}O$ in the stream water (b) from 16:00 of 29th April 2011 to 16:00 of 6th May 2011 storm event.

Figure 4.7 shows the $\delta^{18}O$ values of rainfall calculated using the incremental weighting approach, Eq. (4.5), and the mean values fluctuate between $-1.71‰$ to $-1.48‰$ (Fig. 4.7a). For the three-component hydrograph separation of this event the isotopic signature of rain water (incremental means) was considered (Fig. 4.8). Therefore, the end-member value for rainfall is not constant, but varied over time.

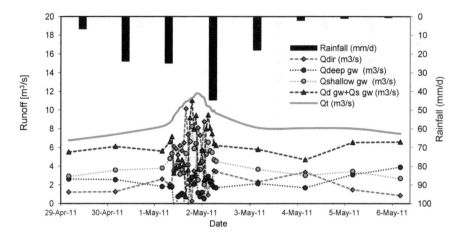

Figure 4.8 Results of the three-component separation using dissolved silica and oxygen-18 as tracers for event M3 (see Fig. 2b) investigated from 29 April 2011 at 16:00 to 6 May 2011 at 16:00 at Migina station. $Q_{dgw} + Q_{sgw}$ is the sum of deep and shallow groundwater components.

Figure 4.8 shows the results of the three-component separation using dissolved silica and oxygen-18 as tracers. During this event, pre-event water (deep and shallow groundwater, $Q_{dgw} + Q_{sgw}$) was chiefly responsible for stream generation and is contributing to the total discharge 10–98% (about 60% on average). Maximum surface runoff generation occurred at the hour of peak discharge (on 2 May 2011 at 10:00 LT) and event water (direct runoff, Q_{dir}) contributes for a short period about 70%. Thus, the peak is dominated by direct runoff but the total discharge (Q_T) is dominated by subsurface water similar to the event of May 2010. However, the results found for this separation are somewhat different from previous results, but the assumptions of the methods are not fully met and cause some uncertainty of the method (Sect. 4.2.3). Unfortunately, there is no independent experimental data that can support the stormflow composition during peak flow.

4.4 DISCUSSION

4.4.1 Rainfall influence on runoff generation

Rainfall and discharge data used in this research were collected over two years (May 2009–June 2011) and the rainy season "Itumba" was investigated in further detail. Low runoff coefficients for different events were determined ranging between 16.7 and 44.5% for Cyihene-Kansi sub-catchment (Table 4.1) and between 31.5 and 44.4% for Migina catchment (Table 4.2). This indicates that the stormflow reaches the stream largely through the subsurface runoff due to high infiltration rates. This type of runoff generation was supported by observed chemical concentrations in river water which are closer to the concentrations of water sampled from springs and piezometers during flood events (Table 4.3).

The high infiltration in the Migina catchment can be explained by a very high hydraulic conductivity of the soil as observed by Van den Berg and Bolt (2010) using double ring infiltrometer tests in the same catchment; the infiltration rate varied between 208 mm h^{-1} to 1250 mm h^{-1}. The tests were conducted at locations where the land is used for agriculture. The rainfall intensities which are less than 17.6 mm h^{-1} are much lower than the infiltration rates (see Tables 4.1 and 4.2). Van den Berg and Bolt (2010) also analyzed maximum soil water content in the soil laboratory and found that the soil can hold up to 60-70% of water. This forms an important shallow subsurface water storage, which makes agriculture possible even in dry periods. Hence, this can lead to a shallow subsurface runoff component contributing to the total streamflow if the storage threshold is exceeded.

Munyaneza *et al.* (2011b) found the long-term average runoff coefficient of Migina catchment to be 25%, which is in the range of the results found in this study. In the same study, they also found that the Migina catchment is dominated by agricultural land use (92.5%). The range of runoff coefficients found in this current study (16.7–44.5%) agrees with the range for agricultural dominated catchments found e.g. by Larsen *et al.* (2007). This gives a hint towards the importance of infiltration and subsurface flow generation during events. Runoff generation obviously depends on other factors such as the degree of slope, soil type, vegetation cover, antecedent soil moisture, rainfall intensity and duration (FAO, 1997). The runoff coefficient ranges usually between 1% and 50% in cultivated catchments (FAO, 2006). Marchi *et al.* (2010) did a study for extreme flash floods in Europe and found that the runoff coefficients of the studied flash floods are usually rather low with a mean value of 0.35. Moderate differences in runoff coefficient are observed between the studied climatic regions, with higher values in the Mediterranean region. Ley *et al.* (2011) found that the annual mean runoff coefficients in nested catchments of Rhineland-Palatinate, Germany, may range from 2% to 15% in the summer period, while during winter time they range from 5% to 56%. The high runoff coefficients observed in Germany in winter are due to snow influence and can be the same as it is the case in the Rwanda areas but it is being due to heavy rainfall observed during the events (see Fig. 4.1). However, the current research was done during the rainy season "Itumba". Therefore, it can be concluded from the rainfall-runoff response analysis that runoff generation at the Cyihene-Kansi and Migina catchments is dominated by subsurface flows as highly supported by the hydrograph separation (see Sect. 4.3.3).

4.4.2 Quantification of runoff components and processes in a meso-scale catchment

Streamflow hydrograph separations were found to be possible using dissolved silica and chloride as tracers due to their variations in concentrations observed during two investigated flood events. However, the remaining analyzed chemical components (SO^{2-}_4, Na^+, K^+, Mg^{2+}, and Ca^{2+}) could not be used for hydrograph separations, because they showed constant concentrations in the streamflow during the events (likely due to non-conservative transport behavior) and did not provide additional insights due to unknown reasons. Their concentrations in surface runoff and groundwater were too similar to do reliable hydrograph separations. Richey *et al.* (1998) used the same method and found that chemical tracers like SiO_2 and Cl^- may be non-conservative in subsurface water on longer timescales, but they can be assumed to behave conservatively on the time scale of a single runoff event. These findings indicate that spatial variability in the components may be more important when determining the precision of the pre-event water fraction. In fact, direct runoff or event water data generated by the selected four tracers in this study offer insights into how the catchments respond hydrologically and were used to develop a conceptual model of how catchment generates runoff.

The two-component hydrograph separation model using dissolved silica and chloride led to a high amount of subsurface contribution (up to 80%) in both catchments. For both investigated events at Kansi and Migina station, the direct runoff component did not exceed 33.7 and 28.7% of the total event runoff, respectively. The observed dominance of subsurface runoff in these two storm events was probably facilitated by the wet conditions during the rainy season (Fig. 4.1) as well as linked probably to soil saturation.

The three-components runoff separation model using dissolved silica and deuterium, and using dissolved silica and oxygen-18 shows somewhat different results but both confirmed the high contribution of pre-event runoff components (about 80% using SiO_2 and ^2H; and about 60% using SiO_2 and ^{18}O). The observed differences could be due to the consideration of spatial and temporal variability of oxygen-18 concentrations in rainfall during the event of May 2011 where rain water was sampled. For the two investigated events (Figs. 4.6 and 4.8), the mean value of the new water component is 31.9 and 38.8% of the total runoff for event of May 2010 and 2011, respectively. The dominance of subsurface water found using three-component separations confirms the findings of temporally highly variable but in total relatively small contribution of surface runoff.

The observed dominance of old water (up to 80%) in the Migina catchment confirms the finding of van den Berg and Bolt (2010) in their study during the dry season. They found that the locations of shallow groundwater in the Migina catchment are between 0.2 m and 2 m, which enables infiltrated rain to reach the groundwater quickly and contribute to subsurface stormflow and later to baseflow. This behavior was explained e.g. by McDonnell (1990) by the fact that the rapid flow of new rainwater during downward and lateral flow in macropores interacts with the soil matrix. The findings of this current paper were also supported by results from several other hydrochemical (and isotopic) studies that found old water and subsurface flow to be the major (more than 50%) component of stormflow in different hydro-climatic rainfall (e.g. Sklash *et al.*, 1976; Sklash and Farvolden, 1979; Kennedy *et al.*, 1986; Rice and Hornberger, 1998; Didszun and Uhlenbrook, 2008; Hrachowitz *et al.*, 2011). Our results are in line with Mul *et al.* (2008) who did a similar study in a semi-arid area using hydrochemical tracers for hydrograph separation and found that over 95% of the discharge could be attributed to subsurface runoff during smaller events, while the remainder was due to faster surface runoff processes. Hrachowitz *et al.* (2011) carried out a study in another semi-arid catchment using hydrometric observation and found that the use of multiple tracers allowed estimating uncertainties in hydrograph separations occurring from the use of different tracers. Applying hydrograph separation methods to larger catchments >40 km^2 often leads to only qualitative results (Uhlenbrook and Hoeg, 2003; Didszun and Uhlenbrook, 2008).

The plateaus of Migina catchment are linked with shallow rivers due to seepage from deeper groundwater towards the river. This can be explained by the temperatures observed in deeper groundwater (through installed piezometers, see Fig. 2.11), which were colder than the shallow groundwater. In contrast the temperatures of river water were colder than shallow groundwater (Van den Berg and Bolt, 2010). Groundwater is deeper in plateaus and the dominated lateral runoff process is Hortonian overland flow (HOF) observed during rainy season events.

At the hillslopes, natural vegetation, forests, and banana trees or small plots with for example sorghum, cassava or maize are planted. The dominant land use of Migina hillslopes is banana trees and other rainfed shrub of herbaceous crops (see Fig. 5.3 and Fig. 4.10). The soils on the hillslopes were affected by weathering processes, which lead to the forming of thick red earths, started to creep from the hills into the valleys. Hillslopes of Migina are affected by erosion due to Hortonian overland flow observed during high rainfall events but the dominated hydrological process is Susurface flow (SSF). Moeyersons (2001) demonstrated that silt layers that intercalate the peat layers are related to slope erosion observed in the Migina hillslopes. Due to the landslide observed, the Government of Rwanda initiated intensive programme of terraces on hillslope areas and the programme is still ongoing. Deep groundwater flow is appearing near plateaus and become shallow groundwater near valleys and generates springs (see Fig. 2.14 and Fig. 4.9) which are located at the contact between the hillslope and the valley floor. The deeper groundwater is separated from shallow groundwater by a impermeable layer which is present in the valleys at 2 to 4 m below the ground surface (Van den Berg and Bolt, 2010).

The valleys in the lower part of the Migina catchment are wider between 50 to 300m. The dominated runoff mechanism is Saturation overland flow (SOF). Closer to the river, the groundwater experiences more influence by the river (Van den Berg and Bolt, 2010). Further away from the rivers, the fields are fed by rain and groundwater.

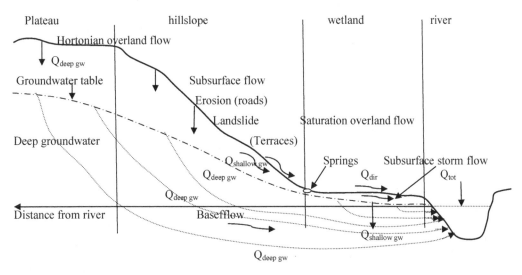

Plateau   hillslope   wetland   river

Hortonian overland flow

$Q_{deep\ gw}$

Groundwater table   Subsurface flow

Erosion (roads)

Landslide   Saturation overland flow

Deep groundwater   (Terraces)

$Q_{shallow\ gw}$ Springs   Subsurface storm flow

$Q_{deep\ gw}$ Q_{dir} Q_{tot}

$Q_{deep\ gw}$

Distance from river   Baseflow

$Q_{shallow\ gw}$

$Q_{deep\ gw}$

Figure 4.9 Migina catchment hydrological processes (not to scale)

Figure 4.10 Photo of the urban area of Butare city in the upstream of Migina (left) and natural vegetation on a hillslope to the west of Butare (right) (Photo taken by Harmen van den Berg, June 2009).

Figure 4.11 Photo of banana trees (left) and irrigated rice fields in the valley to the southwest of Butare, with trees on the hillslope (right) (Photo taken by Harmen van den Berg, June 2009).

Hydrograph separation in the Migina meso-scale catchment helped to gain further insights in runoff process understanding. The runoff components and processes in a meso-scale catchment for two flood events could be estimated and led to a conceptual understanding of the catchment functioning.

4.5 CONCLUSIONS

The applicability of tracer methods in conjunction with hydrometric measurements for identifying dominant runoff generation processes in the meso-scale Migina catchment was tested. The two- and three-components hydrograph separation models using hydrochemical (dissolved silica and chloride) and isotope (deuterium and oxygen-18) tracers show that intensive water sampling (hourly) during events is essential. The whole rising limb, peak and recession limb need to be captured completely for the event in order to gain more understanding of runoff generation processes. In addition, different geographic sources of runoff need to be observed before, during and after the events. The outcomes of such an investigation are essential for understanding the catchment functioning and the sustainable water resources management and agricultural development to meet the high water demands related to the rapid Rwandan population increase.

The results of this study demonstrated the importance of subsurface flows for streamflow generation in the study area. It shows the value of detailed hydrological data collection over two whole rainy seasons using different tracers and hydrometric observation to understand dominant hydrological processes. Furthermore, it demonstrated the significance of considering spatial and temporal variations of rainfall in the hydrograph separations (Figures 4.7 and 4.8); this is of greater importance in meso-scale catchments than in small headwaters. Oxygen-18 (^{18}O) and deuterium (^{2}H) were found to be suitable tracers to detect event vs. pre-event water sources. Additionally, it was found that groundwater has two different origins: one source originates from a near stream location in the valleys (shallow groundwater) and the other source is deep groundwater sampled at piezometers and springs located in the upper part of the hillslopes (Sect. 4.3.3). The significant groundwater recharge during the wet seasons led to the perennial river system observed in the catchment. The isotope analysis showed that all runoff components including baseflow are dependent on wet season rainfall.

Chapter 5

PREDICTION OF RIVER PEAK DISCHARGE IN AN AGRICULTURAL CATCHMENT IN RWANDA

....

Valid criticisms have been raised about the adequacy of the rational method to simulate direct runoff. However, it continues to be used for estimating a design discharge from a small catchment because of its simplicity. The main aim of this work is to contribute to water resources development in a catchment through the development of a simple method to assess the peak flow discharge (10 years return period) of the Migina River (257.4 km²) in Southern Rwanda. A Digital Elevation Model (DEM) with resolution of 90m was used to delineate the catchments. A digitized land use map and hydrological soil group classification of the catchment were used to estimate the runoff coefficient of the sub-catchments to calculate a weighted runoff coefficient. The rainfall intensity was determined using the records obtained from 13 rain gauges installed in the catchment from May 2009 to June 2011 and considering the time of concentration. The catchment profile and empirical formulae were used to compute the time of concentration. A simple rational method with area correction was used to estimate the total discharge from the catchment. The agricultural land use dominates in the catchment (about 92.5%). The results show the weighted runoff coefficient of 0.25, time of concentration of 3.5 hours and the peak flow discharge (10 years return period) of 16 m³ s⁻¹. For assessing runoff and water resources availability on each sub-catchment level,, it is recommended to develop river discharge models using standard methods such as conceptual hydrological modelling to simulate the rainfall runoff of this catchment.

....

Based on: Munyaneza, O., Ufiteyezu, F., Wali, U.G. and Uhlenbrook, S., 2011. *A simple Method to Predict River Flows in the Agricultural Migina Catchment in Rwanda.* Nile Water Sci. Eng. J., 4(2): 24-36.

5.1. INTRODUCTION

The Rwandan economy is rapidly developing with a rate of 5% and more in the last years. However, over 85% of the population depends on subsistence agriculture for their livelihoods. This makes the issue of water resources availability very crucial. So far, the water resources of the country have not been quantified and there are no reliable data on the amount of water resources.

Accurate and comprehensive knowledge about water resources form the basis for effective water resources management. It is now widely recognized that the monitoring and assessment of water resources, in terms of both quantity and quality, require adequate hydrological and meteorological data (Maathius *et al.*, 2006), which is often a challenge due to the spatio-temporal variation of processes (Uhlenbrook, 2006). Wagener *et al.* (2008) stated that about one billion people live in water-scarce or water-stressed regions, and by 2025 this number is expected to increase further significantly. The magnitude of this water scarcity and its variations in both space and time are largely unknown and all estimations are uncertain, and one of the main reasons is the lack of hydro-climatological data (Oyebande, 2001; Kipkemboi, 2005).

In Rwanda, the rapid population growth is increasing the competition for the available water resources amongst domestic, agricultural, industrial and other user, thereby increasing water scarcity. In addition, the limitation of knowledge in water resources assessment and management is making the problem of water scarcity in Rwanda more pronounced. This has a negative impact on the agriculture development and the country's economy is mostly based on agriculture (Nahayo *et al.*, 2010).

As a solution for food security and poverty alleviation problems, Rwandan marshlands are being developed for intensive agricultural activities (World Bank, 2008) and the implementation of the Rwanda irrigation master plan by the Ministry of Agriculture have commenced. To achieve efficient implementation of these activities would be difficult without knowing the water resources availability in the catchment (Munyaneza *et al.*, 2010). Therefore, it is very important to improve the understanding of different components of the hydrological cycle and the spatial and temporal distribution of water at present and in the future so as to improve the management of water resources.

The rational method is a simple technique for estimating a design discharge from a small catchment (Thompson, 2006). The rational method, which can be traced back to the middle of the 19th century, is still probably the most widely used method in applied hydrology for small catchments. The Irish engineer, Mulvaney (1850), was the first to publish the principles on which the method is based, although Americans tend to credit Kuichling (1889) and the British Lloyd Davies (1906) for the method (Viessman *et al.*, 1989). It is an empirically developed model, with simplifying assumptions including uniform rainfall with uniform intensity over the entire watershed for duration to the time of concentration (Haan *et al.*, 1982). The method has been adapted to Rwandan conditions and checked with measurements as much as possible.

The time of concentration, t_c, of a catchment is defined to be the time required for a parcel of runoff to travel from the most distant part of a catchment to the outlet (Thompson, 2006). The concept of time of concentration is a rough simplification of a complex reality; however, it has been proofed useful in practice for describing the time response of a catchment to an impulse. In the context of the rational method, time of concentration represents the time at which all areas of the catchment that generate runoff are contributing runoff to the outlet.

The objective of this paper is to introduce a simple method to estimate peak flow in a typical agricultural catchment located in Southern Rwanda. Therefore, different land cover, land use and soil type in the Migina catchment are classified; the runoff coefficients within Migina catchment through regrouping of land cover/land use and hydrological soil group maps are determined; and a simple method for maximum peak flow estimation is developed.

5.2. DATA COLLECTION AND PROCESSING TECHNIQUES

5.2.1. Physical characteristics of the catchment

The topography of the Migina catchment was extracted from the Digital Elevation Model (DEM) obtained from the USGS website[11] with the resolution of 90 m. The topography was described in terms of slope, aspect, and elevation. These three basic characteristics affect the movement and storage of water in the catchment. The catchment boundary was delineated from the DEM within the ArcGIS environment. The land cover map of Rwanda was used to extract the Migina catchment's land cover. The map was collected from the Center of GIS at Butare and was taken in 2009.

Hydrological soil groups: The type of soil in a catchment plays an important role in the catchment's hydrology and in particular in the runoff generation processes. The composition and texture of the soil determines whether rainfall or irrigation water will be retained in the soil and released gradually, percolating downward to the groundwater, or if these inputs of water instead contribute to surface runoff, leading to increased erosion. The hydrological soil group (HSG) of the study area was extracted from a map the soil texture classes of the dominant soil units in Rwanda in the ArcGIS environment (source: Verdoodt and van Ranst, 2003).

5.2.2. Determination of Rainfall Intensity

The catchment is equipped with meteorological instruments include: two weather stations that record rainfall, temperature, relative humidity, soil moisture, wind speed and direction. Those were installed downstream at the Gisunzu Primary School and in an upstream area at the CGIS Center in the city of Butare (Fig. 2.4). Three tipping bucket gauges were installed in the western, eastern and central part of the catchment. These locations were selected because of the other tipping buckets installed at the weather stations and to have a regional spread (Fig. 2.4). However, data from these tipping buckets could not be used in this paper to estimate rainfall intensity because of technical problems and observer problems. Many missing data sets were observed and rainfall data collected from raingauges were used in this paper.

The rainfall intensity, *I [mm hr^{-1}]*, is the average rainfall rate in mm hr^{-1} for a particular drainage basin or sub-basin. The intensity should be selected on the basis of the design storm duration and return period (Dawod *et al.,* 2011). The design storm duration is equal to the time of concentration for drainage area under consideration. The return period is established by design standards or chosen by the hydrologist as a design parameter.

Deciding upon an adopted return period for the design storm should be based on cost benefit analysis, which quantifies the physical and social damage caused by flooding. However, no conclusive advice appears to have been made which quantifies the actual costs of flooding. Thus, the selection is normally based on local experience. In many cases, the hydrologist has a standard IDF (Intensity-Duration-Frequency) curve or table available for the site and does not have to perform this analysis even though he should (Butler and Davies, 2004). The intensity-duration-frequency curves (IDF curves) represent the relation between three components, storm duration, storm intensity, and storm return interval (Thompson, 2006).

The average of rainfall intensity in Migina catchment was estimated after calculating the daily rainfall by the use of an arithmetic mean method (Eq. 5.1) and considering the time of concentration, *t$_c$ [min]*, (Dawod *et al.,* 2011). Data used to assess the rainfall intensity were collected from 13 rain gauges in a two-year period on a daily basis (May 2009-June 2011). The arithmetic mean of the rainfall amounts measured in the area provides a satisfactory estimate for a relatively uniform rain.

---

[11] http://www.dgadv.com/srtm30/

$$P = \frac{P_1 + P_2 + P_3 + \cdots + P_n}{n} \qquad (5.1)$$

where: P = Precipitation (mm/d), and n = total number of rain gauges (-).

Alternatively, rainfall intensity can be computed by the recommended standard IDF equations (Butler and Davies, 2004), and the Equation (5.2) was selected due to the field investigation on the decade floods occurred in the Kagera River basin as monitored in 2009 by Munyaneza *et al.* (2011a). Based on their preliminary results, they have assumed that the flood frequency in the Kagera River basin is 10 years, but they suggested that an advanced study is needed to check or confirm this assumption based on longer rainfall time series available all over the basin. Note, that Migina catchment is a sub-catchment of Kagera River basin.

$$I_{10} = \frac{140}{t_c + 0.7} \qquad (5.2)$$

where I_{10} is intensity in (mm hr^{-1}) for the return period of 10 years and t_c is time of concentration in mins.

The applied method assumes that the maximum runoff rate in a catchment is reached when all parts of the catchment are contributing to the outflow. This happens when the time of concentration is reached. The Kirpich/Ramser formula (Kirpich, 1940) is mostly used to calculate the time of concentration (De Brouwer, 1997; De Laat and Savenije, 2002; Dawod *et al.*, 2011):

$$t_c = 0.0195 L^{0.77} S^{-0.385} \qquad (5.3)$$

where: t_c = Time of concentration [min], L = Maximum length of the catchment [m], S = the slope of the catchment over the distance L [m/m]. In cases there are no upstream drains, the inlet time is considered as the time of concentration and is usually obtained by the Kirpich (Kirpich, 1940) or the Lloyd-Davies (not used in this study) formula which include the time of entry $t_e = \dfrac{58.5L}{A^{0.1} S^{0.2}}$. All the above formulae are very simple empirical equations that are not fully physically based and the units do not add up.

5.2.3. Estimation of runoff coefficient

The runoff coefficient (C) is the least precise variable of the rational method ($0 < C \le 1$, [-]) (Mihalik, 2007). It is a dimensionless ratio intended to indicate the amount of runoff generated by catchment given an average intensity of precipitation for a storm (Thompson, 2006). Its use in the formula implies a fixed ratio of peak runoff rate to rainfall rate for the drainage basin, which in reality is not the case. However, the runoff coefficient [-] from an individual rainstorm is normally defined (Hawkins, 1975) as runoff divided by the corresponding rainfall, both expressed as depth over catchment area in mm:

$$C = \frac{Q}{P} \qquad (5.4)$$

All catchment losses and water storages are incorporated into the runoff coefficient, which is usually a function of the land use, soils, geology, etc. This means that the magnitude of this coefficient is not constant, but varies with time and in space depending on a number of different factors such as the topography of the catchment, magnitude and intensity of the storm rainfall, vegetation cover and land use, infiltration rate and the initial soil moisture condition, groundwater depths and subsurface flows (e.g. EFM, 1984 in Musoni *et al.*, 2010; Uhlenbrook 2007). All of these factors influence the hydrological connectivity.

The Tables 5.1, 5.2 and 5.3 describe the Migina land cover, the runoff coefficient for agricultural watersheds and the hydrologic soil group conversion factors. These three tables were used to estimate the runoff coefficient of the Migina catchment. This is due to the fact that the runoff coefficient is dependent on the character and condition of the soil. The use of Tables 5.2 and 5.3 was supported by De Brouwer (1997). He demonstrated that those tables can be used in agricultural catchments to determine the peak runoff rate when applying the rational formula which is an empirical equation. This was also supported by Verdoodt and van Ranst (2003) who conducted studies in agricultural catchments in Rwanda.

Table 5.1: Migina land cover description (based on Verdoodt and van Ranst, 2003).

Land cover ID	Land cover name	Land use/ Land cover
AG-10*/5	Combination of Banana Plantation and Rain fed Herbaceous Crop	Agriculture: Small grain, good practice
AG-2	Irrigated Herbaceous Crop	
AG-5/9	Combination of Rain fed Herbaceous Crop-Two crop per year and Shrub Plantation	
AG-9/5	Combination of Shrub Plantation and Rain fed Herbaceous Crop-Two crop per year	
AG-6	Forest Plantation - (Eucalyptus or Pinus and Cypress)	Forest: Woodland, mature, good
AG-6B	Scattered (in natural vegetation or other) Forest Plantation (Eucalyptus or Pinus and Cypress)	
AG-6C	Isolated (in natural vegetation or other) Forest Plantation (Eucalyptus or Pinus and Cypress)	
RL-2	Savannah (shrub or tree and shrub)	Grass/Lawn: Pasture, permanent, good
UR	Urban And Associated Areas	Building

Table 5.2: Runoff coefficient C for agricultural watersheds (soil group B) (based on Schwab et al., 1993)

Crop and hydrologic condition	Coefficient C (-) for rainfall rates of		
	25 mm hr^{-1*}	100 mm hr^{-1*}	200 mm hr^{-1*}
Row crop, poor practice	0.63	0.65	0.66
Row crop, good practice	0.47	0.56	0.62
Small grain, poor practice	0.38	0.38	0.38
Small grain, good practice	0.18	0.21	0.22
Meadow, rotation, good	0.29	0.36	0.39
Pasture, permanent, good	0.02	0.17	0.23
Woodland, mature, good	0.02	0.10	0.15

means that these intensities are super high and normally do not occur.

Table 5.3: Hydrologic soil group conversion factors (based on Schwab et al., 1993)

Crop and hydrologic condition	Factors for converting the runoff coefficient C from group B soils to		
	Group A	Group C	Group D
Row crop, poor practice	0.89	1.09	1.12
Row crop, good practice	0.86	1.09	1.14
Small grain, poor practice	0.86	1.11	1.16
Small grain, good practice	0.84	1.11	1.16
Meadow, rotation, good	0.81	1.13	1.18
Pasture, permanent, good	0.64	1.21	1.31
Woodland, mature, good	0.45	1.27	1.40

In this paper, the runoff coefficient of the Migina catchment was estimated after making an overlay of a land cover map and hydrological soil group map using ArcGIS tools. A weighted average runoff coefficient was calculated using Equation (5.5) as modified from Chow et al. (1988) to be applicable to larger watersheds.

$$C = \sum_{n=1}^{n} \left(\frac{A_i C_i}{\sum_1^n A_i} \right) \tag{5.5}$$

Where: A_i (km²) represents partial areas and C_i [-] the partial runoff coefficients, according to five sub-catchments of Migina catchment.

5.2.4. Determination of Peak Runoff

The rational formula is a simple method, but widely used in applied hydrology for small catchments. The method has been adapted to Rwandan conditions and checked with measurements as much as possible (Verdoodt and van Ranst, 2003; Munyaneza et al., 2012a). After estimating the rainfall intensity of the Migina catchment, information on Migina catchment describing topographical configuration and the runoff coefficient from land cover and Hydrological Soil Group (HSG), the peak runoff was computed using the following form of the rational formula (Haan et al., 1982; De Laat and Savenije, 2002) for the return period of 10 years.

$$Q_p = CIA \tag{5.6}$$

where: Q_p = Peak runoff rate [m³ s⁻¹], C = Runoff coefficient [-] (dependent on catchment characteristics), I = Intensity of rainfall during time t_c [m s⁻¹], and A = Catchment area [m²]. The value of I is assumed constant during t_c and the rain uniformly distributed over A. The peak flow Q_p occurs at time t_c).

The idea behind the rational method is that if a rainfall of intensity I [mm] begins instantaneously and continues, the rate of runoff will increase until the time of concentration t_c [min], which means the time at almost all of the watershed is contributing to flow at the outlet. The duration used for determination of the design storm intensity is the time of concentration of the watershed. The rational method was applied in this paper by considering the following assumptions (Haan et al., 1982):

1. Rainfall intensity is constant throughout the storm duration.
2. The time of concentration employed is the time for the runoff to become established and flow from the most remote part of the drainage area to the inflow point of the drain to be designed (Fig. 5.2).
3. The computed peak rate of runoff at the outlet point is a function of the average rainfall (I) rate during the time of concentration, that means the peak discharge does not result from a more intense storm of shorter duration, during which only a portion of the watershed is contributing to runoff at the outlet.
4. It has been found by experiments that as the catchment area increases (>2 km²) the rational formula becomes less accurate (Langousis, 2005). De Laat and Savenije (2002) recommended its application in small catchments (smaller than 15 km²), but it was applied in the meso-scale Migina catchment for testing its applicability. In such a case the point area should be multiplied by ARF (Area Reduction Factor) as shown in Fig. 5.1. The application of ARF to larger catchments was supported by Chow et al. (1988) and its accuracy was tested successfully by Butler and Davies (2004). Butler and Davies (2004) said "Point rainfall is not

necessarily representative of rainfall over a larger area because average rainfall intensity decreases with increasing area. In order to deal with this problem, and avoid overestimating flows from larger catchments, *areal reduction factors* (ARF [-]) have been demonstrated by Chow *et al.* (1988). The expression is valid depending on climate and/or storm duration.

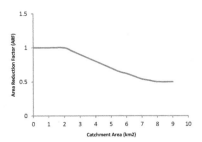

Figure 5.1: *Area reduction factor [-] curve for larger catchments (adapted from Langousis, 2005).*

The rational formula was developed in an area like this for the estimation of peak discharge in small catchments, but Chow *et al.* (1988) demonstrated that it can also be applicable in meso-scale catchments. Therefore, it was applied in the Migina catchment. Many methods for estimating runoff exist (e.g. Haan *et al.*, 1982; Chow *et al.*, 1988), and rational method was used in this study and checked with field observations for its applicability (see Sect. 4.3.1).

5.3. RESULTS AND DISCUSSIONS

5.3.1. Data Processing and Results

5.3.1.1 Catchment delineation

Figure 5.2 shows topographic character of the Migina catchment that is very mountainous with elevation ranging from 1,375 m, where the Migina River is entering the Akanyaru River, to 2,278 m at Huye Mountain (North West). The longitudinal slopes of the valleys vary from 5 to 10% upstream and from 1 to 15% downstream (average slope is between 2 and 3%). This has been also reported by Nahayo *et al.* (2010).

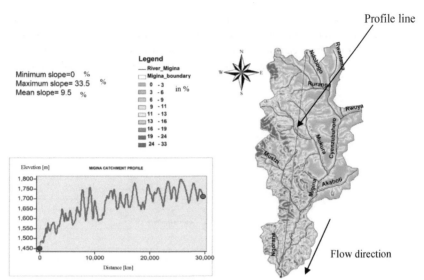

Figure 5.2: The map shows the Migina catchment slope gradient and the inset figure displays the catchment profile.

The average slope of main stream from a profile in the Figure 5.2 is 9.5%. The profile was plotted a long one line as shown in Figure 5.2. The total area, which is 257.4 km², has been calculated by use of Arc map geometry analysis. The main stream length is around 40.0 km (Fig. 5.2). The map used in this paper has a resolution of 90 m. High resolution satellite images are not easily available and hence lower resolutions images can be applied (Munyaneza *et al.,* 2009a), which could affect the accuracy during the computation of the slopes.

5.3.1.2. Calculation of time of concentration
The Kirpich/Ramser formula (Eq. 5.3) has been used to calculate the time of concentration of the Migina catchment considering the length of the main river and the average slope of main stream, and was found to be 3 hrs 26 min.

5.3.1.3. Calculation of rainfall intensity
Average rainfall intensity was estimated using average (arithmetic) method (Eq. 5.1) from the daily rainfall data taken during 2 years (May 2009-June 2011) and by considering the time of concentration of 3 hours 30 mins. The estimated average rainfall intensity was estimated as 1.06 mm hr⁻¹. Note, this might appear low but daily data had to be used and the event duration is usually shorter, and the model assumes spatially uniform rainfall what is also not the case in tropical environments. Then, the empirical Equation 5.2 was also used for double-checking and the rainfall intensity was found to be 0.68 mm h⁻¹. After the application of both methods, the higher rainfall intensity was used to predict the Migina River peak flows for the security of future hydraulic structures design.

5.3.1.4. Determination of the Migina runoff coefficient

Computation of Migina land cover
Figure 5.3 shows that the land cover of Migina catchment is composed of the followings types: Agriculture (small grain, good practice), Forest (woodland, mature, good), Grass/lawn (pasture, permanent, good), and Urban areas as also found by Verdoodt and van Ranst in 2003. The only

difference found in this paper is the percentage of covering areas. The catchment land use is widely dominated by agriculture activities at 92.5% on D (sandy 50%), C (loamy 47%) and B (clayey 3%) soil groups; forest at 5% on C (loamy 15.5%), D (sandy 37.3%) and B (clayey 47%) soil groups; grass/lawn at 2% on C (sandy 100%) soil group and, finally, buildings which cover 0.5% on D (loamy 100%) soil group. Based on field information, the catchment land use is dominated by pasture and arable farming like rice, sorghum, maize and sweet potatoes (Munyaneza *et al.*, 2010).

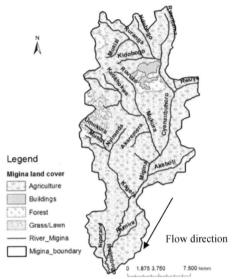

Figure 5.3: Land cover of the Migina catchment.

Classification of hydrological soil groups of the Migina catchment
Figure 5.4 shows that Migina catchment is dominated by group D and C which means very high runoff potential and moderate to high runoff potential, respectively. Group A which represents low runoff potential is not available in the Migina catchment. Only the group B of low to moderate runoff potential was found (Fig. 5.4).

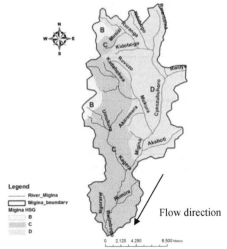

Figure 5.4: Hydrological soil groups of the Migina catchment.

Overlay of land cover and hydrological soil groups maps of the Migina catchment

Figure 5.5 was computed by overlaying the land cover and hydrological soil groups of the Migina catchment. The information from the figure was used in Table 5.4 to estimate the runoff coefficient of the Migina catchment.

Figure 5.5: Overlay of land cover and hydrological soil groups of the Migina catchment.

The Migina catchment runoff coefficient was calculated using Eq. 5.5 (Table 5.4) and the runoff coefficient was obtained after an overlay of a land cover map and soil type map (Fig. 5.5). It has been found to be around 0.25 (Table 5.4).

Table 5.4: Summary of determination of runoff coefficient

Land cover	Hydrological soil group	Area (A_i) km^2	Runoff coefficient (C_i)	(A_i) × (C_i) km^2
Agriculture: Small grain, good practice	D	119.92	0.24	29.21
	C	111.61	0.23	26.02
	B	6.81	0.21	1.43
Forest: Woodland, mature, good	C	2.14	0.127	0.27
	D	5.23	0.14	0.73
	B	6.51	0.10	0.65
Buildings	D	4.94	1.00	4.94
Grass/ Lawn: Pasture, permanent, good	C	0.28	0.242	0.07
Σ		257.43		63.32

$$C = \sum_{n=1}^{n} \left(\frac{A_i \times C_i}{\sum_1^n A_i} \right) = \frac{63.32}{257.43} = 0.2459 \cong 0.25$$

The values of C_i were obtained by the use of Tables 5.1, 5.2 and 5.3.

5.3.1.5. Determination of the Migina peak runoff

The Migina catchment peak runoff was calculated using the Equation 5.6 by considering the area reduction factor (Fig. 5.1) and was found to be 16.12 m^3 s^{-1} (62.6 l km^{-2} s^{-1}).

5.3.2. Discussion

5.3.2.1. Migina River discharge prediction

In order to establish a relationship between water levels and discharge for the five main streams in the Migina catchment, several discharge measurements were carried out at the gauging stations at different water levels (Munyaneza et al., 2010). The rating curve was generated according to the recommendations of Shaw (2004) as shown in Equation (2.1) and the Migina mean daily discharges were also discussed in Munyaneza et al. (2010) after considering field data observed from all stations.

Average rainfall intensity was estimated using average (arithmetic) method and considering the time of concentration of 3 hours and 30 mins and is around 1.06 mm hr^{-1} while the total catchment area is equal to 257.4 km^2 (Fig. 5.2). We have double-checked these results using previous researches conducted in the Migina catchment (e.g. SHER, 2003; Nahayo et al., 2010; van den Berg and Bolt, 2010; Munyaneza et al., 2010) and we can stand behind them, however, the used rainfall data are still doubtful because a 10-year rainfall event should have an intensity of several tens of mm per hour. A one mm per hour event is very low and can not produce a flood. The reason is that in reality, the rainfall is not distributed uniformly over the catchment, but spatially highly variable and locally much higher rainfall intensities are reached. However, a lumped model can not deal with complexity, therefore, through the parameterisation the model 'gets right for wrong reason', i.e. accounting for neglecting rainfall variability through low rainfall intensities that are assumed uniform over the catchment. The high time of concentration value found in the Migina catchment has a big influence on the low rainfall intensity. The low runoff coefficient computed in the same catchment which is around 0.25 can even be motivated (Table 5.4), due to agriculture activities carried out in the studied Migina Catchment. This coefficient found in the Migina catchment agrees with the range for an agricultural dominated catchment as found by Purdue et al. (2007).

The Migina catchment was considered to be large catchment for the application of rational method (meso-scale catchment, almost 260 km^2), though the correction factor ARF was applied (Fig. 5.1), and the maximum peak flow was predicted to be 16 m^3 s^{-1}. The agriculture activities in wetlands contributed significantly for the computation of this peak runoff (Table 5.4). These results are not very different from the results obtained based on rainfall-runoff observations in the same catchment. Munyaneza et al. (2012a) observed discharges at the outlet of the Migina catchment (at Migina station) during the period 1st August 2009 to 31st June 2011. They found that streamflows in the Migina catchment range between 0.43–15.60 m^3 s^{-1}, including one large event. This makes the findings of this study plausible (see Sect. 4.3.1). Based on this comparison, we recommend that this method of river peak flow prediction can be applied carefully to other agricultural catchments in the region, but simple hydrological studies have to be done for getting the required data to plot runoff and rainfall intensity relationship curves. Rainfall runoff hydrological models are recommended (e.g. HBV, HEC-HMS, FLEX, etc.). The semi-distributed conceptual hydrological catchment model HEC-HMS was used in this study due to this recommendation (see Chap. 6).

5.3.2.2. Challenges

Challenges related to urbanization
The country of Rwanda is quickly progressing in terms of development; this implies a change from rural areas to urban areas. Urbanization can have great impacts on the hydrological cycle. It involves mainly increasing overland flow. These artificial surfaces increase significantly the amount of runoff in relation to infiltration, evaporation and interception (total evaporation) and, therefore, increase the total volume of water reaching the river during or soon after the rain.

Challenges related to baseflow contribution
Once groundwater table is not far from the ground surface, as it is the case in many areas of the Migina catchment (often between 0.8 m and 2 m or even less, in particular in the valleys (van den Berg and Bolt, 2010), part of the infiltrated rain can reach the groundwater quickly and contribute to subsurface storm flow and baseflow as discussed in Chapter 6 of this study using HEC-HMS model. Unfortunately, in the suggested simple runoff estimation method, the groundwater contribution during an event was assumed constant. The rational method only simulates "direct runoff", which is in reality a combination of Horton overland flow, saturated overland flow and subsurface storm flow. This needs to be addressed in future developments to improve the method further. E.g. the Hydrologic Engineering Center's Hydrologic Modelling System (HEC-HMS) calculates outflow from the sub-catchment element by subtracting rainfall losses (i.e. total evaporation and retention of event water), calculating surface/direct runoff and adding baseflow (USACE-HEC, 2010).

5.4. CONCLUDING REMARKS
The Migina catchment is a mountainous watershed with elevation ranging from 1,375 m a.s.l. at the outlet to 2,278 m a.s.l. at Mount Huye, which is located in the north-western part of the catchment (Nahayo *et al.*, 2010). It has an area of around 257.4 km^2 and an average runoff coefficient of 0.25. This value falls in the range for an agricultural dominated catchment as found by Purdue *et al.* (2007). With the suggested simple empirical approach and by using data from field investigations, the Migina catchment peak runoff discharge could be predicted. The river peak discharge generated in the Migina catchment on average every 10 years is around 16 m^3 s^{-1} (62.6 l km^{-2} s^{-1}) by use of an area reduction factor and the rainfall intensity observed in the Migina catchment. Results found by Munyaneza *et al.* (2012a) showed that the maximum discharge in the Migina catchment was up to 15.6 m^3 s^{-1} using discharge data measured from 1st August 2009 to 31st June 2011, including one large event. The time of concentration was estimated to 3.5 hours. Land cover and hydrological soil groups analyses in the Migina catchment show that it is dominated by agriculture activities (92.5%) while forest occupy 5%; grass/lawn 2% and buildings cover 0.5%.

It is recommended that this approach is applied to similar agricultural dominated catchments in the region and results could be used in flood management and support decision making. This recommendation is based on rational method plausibility check which has been performed using field observations (Verdoodt and van Ranst, 2003, Nahayo, 2008, Nahayo et al., 2010, Munyaneza et al., 2012a)).

For assessing runoff and water resources availability on sub-catchment level,, it is recommended to develop better process-based river discharge models in the future using standard methods such as hydrological modelling to simulate continuously the rainfall runoff of this catchment (see Chap. 6).

Chapter 6

ASSESSMENT OF SURFACE WATER RESOURCES AVAILABILITY USING CATCHMENT MODELLING AND THE RESULTS OF TRACER STUDIES

In the present study, we developed a hydrological model of the catchment, which can be used to inform water resources planning and decision making for better use of Migina catchment (257.4 km^2). The semi-distributed hydrological model HEC-HMS (version 3.5) was used with its soil moisture accounting, unit hydrograph, liner reservoir (for baseflow) and Muskingum-Cunge (river routing) methods. We used rainfall data from 12 stations and streamflow data from 5 stations, which were collected as part of this study over a period of two years (May 2009 and June 2011). The catchment was divided into five sub-catchments. The model parameters were calibrated separately for each sub-catchment using the observed streamflow data. Calibration results obtained were found acceptable at four stations with a Nash–Sutcliffe Model Efficiency index (NS) of 0.65 on daily runoff at the catchment outlet. Due to the lack of sufficient and reliable data for longer periods, a model validation (e.g. split sample test) was not undertaken. However, we used results from tracer based hydrograph separation to compare our model results in terms of the runoff components. The model performed reasonably well in simulating the total flow volume, peak flow and timing as well as the portion of direct runoff and baseflow. We observed considerable disparities in the parameters (e.g. groundwater storage) and runoff components across the five sub-catchments, which brought better understanding into the different hydrological processes at sub-catchment scale. We conclude that such disparities justify the need to consider catchment subdivisions, if such parameters and components of the water cycle are to form the base for decision making in water resources planning in the catchment.

Based on: Munyaneza, O., Mukubwa, A., Maskey, S., Wenninger, J.and Uhlenbrook, S., 2013. *Assessment of surface water resources availability using catchment modelling and the results of tracer studies in the meso-scale Migina Catchment, Rwanda.* Hydrol. Earth Syst. Sci. Discuss., 10: 15375-15408.

6.1. INTRODUCTION

Water resources availability is often the most vital factor controlling the economic growth in developing countries, which depend on agriculture (Abushandi, 2011). It is obvious that the water challenges will be of utmost and increasing significance throughout the next decades. Extensive care should therefore be given to the operation and management of river basins, focusing on water supply, irrigation, and drought or flood control, in order to cope with water related problems. This situation also applies to Rwanda, where the implementation of sustainable water management interventions is essential to increase or sustain water resources, especially for the agriculture and livestock sectors (UNEP, 2005). The same situation drove the Rwandan government to implement new projects that provide the country with more usable fresh water and increase water availability in the marshlands for agricultural purpose (MINITERE, 2005). Unfortunately, the farmers who use these marshlands do not have appropriate methods for maximizing their production due to the lack of knowledge on water availability in the marshlands. Water resources assessment at the catchment scale is therefore one of the key activities to provide insight on water available for agricultural purpose (Abdulla et al. 2002; Al-Adamat et al., 2010).

The water resources availability assessment requires detailed insights into hydrological processes. However, studying the complexity of hydrological processes, needed for sustainable catchment management, is basically based on understanding rainfall characteristics and catchment properties (Abushandi, 2011), which calls for rainfall-runoff modelling studies (Yener et al., 2007). Rainfall-runoff models have been broadly used in hydrology over the last century for a number of applications, and play an important role in optimal planning and management of water resources in catchments (e.g. Pilgrim et al., 1988; O'Loughlin et al., 1996). Pilgrim et al. (1988) and Oyebande (2001) reported that the main challenge associated with applying successfully rainfall-runoff model lies in the lack of monitored data, mainly rainfall spatial distribution over the catchment area, since rainfall is the primary input in any hydrological model. Another potential problem is having no reliable flow data that can lead to reliable calibration and validation of catchment parameters. In particular, the latter challenge applies to Rwanda, where many catchments are ungauged or even those gauged have unreliable information.

In the last five years, Rwanda has been moving from centralized to decentralized water resources management. This has been done in line with addressing the goal number 7 (to ensure environmental sustainability) of the Millennium Development Goals (MDGs) by elaborating Rwanda Vision 2020, EDPRS I (2007-2012) and EDPRS II (2013-2018). The ultimate goal is to manage water resources in an integrated way and at the lowest possible basin level. The Rwanda National Water Resources Master Plan (RNRA, 2013) has divided the country's watershed into four levels with two main basins on the first order (Congo and Nile). The Migina catchment falls under the third level, within catchments have more or less uniform hydrological characteristics (mostly defined by land use, topography and geology). The surface areas of basins of the third level are typically of the order of at least 10 to possibly some hundreds of km² (RNRA, 2013), and it is at that level that all water resources interventions shall be planned. In other words, for sustainable water resources planning and management, development and related environmental interventions shall be tailored to the characteristics of a specific catchment. Therefore, not only the findings of this study will contribute to enhance the knowledge base, but will also contribute to informed decision making in water resources development planning in the Migina catchment.

A number of research studies have been conducted in this catchment during the last few years (SHER, 2003; Nahayo et al., 2010; van den Berg and Bolt, 2010; Munyaneza et al., 2010, 2011b, 2012a,b). Munyaneza et al. (2012a) applied the two-component hydrograph separation model in two sub-catchments of Cyihene-Kansi and Migina using dissolved silica (SiO_2) and chloride (Cl^-) as tracers determining the contributions of direct and base flows to the total outflows from the two sub-catchments. Munyaneza *et al.* (2012a) also used the three-component runoff separation model with dissolved silica and deuterium, and dissolved silica and oxygen-18. They demonstrated the importance of subsurface flow components (i.e. shallow and deep groundwater runoff) in the Migina catchment.

The University of Rwanda (UR), Huye Campus, which lies in the Migina catchment, supported the idea of building a pilot demonstration site on which models can be built, tested, and results integrated in water resources development planning processes. The approach applied on the Migina can be used for similar studies in other catchments in the region.

In the present study, the Hydrologic Engineering Center-the Hydrologic Modelling System (HEC-HMS) was adopted as hydrologic modelling tool for assessing the water resources availability in a meso-scale catchment, due to its simplicity in setting-up, low data demand for running simulations, and the fact that it is a public domain software.

The HEC-HMS is a semi-distributed conceptual hydrological model, designed to simulate the rainfall-runoff processes for catchment systems (USACE, 2008, Scharffenberg and Fleming, 2010). Its design allows applicability in a wide range of geographic areas for solving diverse problems including large river basin water supply and flood hydrology, and small urban or natural catchment runoff (Merwade, 2007). The model contains parameters that cannot frequently be measured directly, but can only be estimated by calibration using historical records of measured input and output data. The simulation results, especially the water balance components, provide information on water resources available in a catchment for different purposes including, but not limited to, agriculture and domestic purposes. The flow results coupled with the basin characteristics (slopes and imperviousness) can also be used in planning for watershed management measures including but not limited to erosion control, soil moisture and land management measures (Sardoii *et al.*, 2012).

Many researchers have used the rainfall-runoff simulation methods contained in HEC-HMS (e.g. and Yung, 2001; Emerson *et al.*, 2003; Radmanesh *et al.*, 2006; Sardoii *et al.*, 2012). For instance, Radmanesh *et al.* (2006) calibrated and validated the HEC-HMS model in a catchment using different methods incorporated in the model. Their results showed that the SCS (Soil Conservative Service) method resulted in better agreement between peak discharge of observed and simulated hydrographs than other HEC runoff computation methods. Rainfall-runoff correlation in HEC-HMS was modeled by Emerson *et al.* (2003). Results revealed that natural reserved and protected areas decrease the peak of storm events. Christopher and Yung (2001) carried out a study by using HEC-GeoHMS and HEC-HMS to perform a grid-based hydrologic analysis of a catchment. They compared distributed, semi-distributed and lumped models. The results showed sound predictions to observations of flood and runoff volume. The main objective of this study is to analyse spatial variation of runoff generation characteristics of the Migina catchment using a semi-distributed hydrological model with a view to potentially use it for informing water resources planning and decision making. The model is calibrated using detailed two years of rainfall and runoff data collected as part of this study and tracer-based hydrograph separation results (Munyaneza *et al.*, 2012a) are used for validation of the model in terms of runoff components.

6.2. DATA AND METHODS

The assessment involved collecting and screening required data, selecting and building the rainfall-runoff model, calibrating the simulated flows for each individual sub-catchment, and analyzing and interpreting the results.

6.2.1. Data

In order to build the model, the following meteorological and hydrological data were collected: (i) rainfall; (ii) temperature; (iii) solar radiation; (iv) relative humidity; and (v) stream flows. In the framework of this work, the Migina catchment was equipped with 13 and 5 stations rainfall and streamflow instruments, respectively. Rainfall and runoff data were collected over two years (May 2009 to June 2011), whereas other meteorological data were obtained from the CGIS station (Butare), which is operational since February 2006. Rainfall data from 12 stations were only used in this study, given that the rainfall data collected at the CGIS station were not complete. The water levels were measured continuously at five river gauging stations using manual recorders (staff gauges) and pressure transducers (Mini-diver). Rating curves were established using discharge measurements at different periods from May 2009 to June 2011 (Eq. 2.1). The recorded water levels were converted into discharge values using rating curves ($r^2 = 0.88$, $n = 25$ at Rwabuye station; $r^2 = 0.96$, $n = 25$ at Akagera station; $r^2 = 0.94$, $n = 24$ at Kansi station; $r^2 = 0.80$, $n = 28$ at Mukura station; and $r^2 = 0.97$, $n = 18$ at Migina station). Daily temperature and solar radiation data used to compute evaporation were collected at the CGIS-Meteo station using Priestley-Taylor method on daily data basis.

Rainfall data at 12 stations scattered in the study area were analysed using Thiessen polygon method for the interpolation of the daily rainfall average and using the Mass Curve Method for quality control as shown in Figure 6.1.

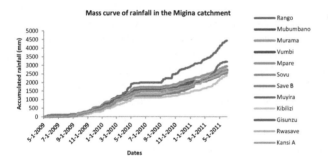

Figure 6.1 Mass curve of rainfall at 12 stations around the Migina catchment for the period of May 2009 to June 2011.

Figure 6.1 shows that all plotted mass curves of rainfall in the Migina catchment have similar behavior except for Rango station which shows significantly higher rainfall than other stations due to unknown reasons. The station was still used in the analysis as there was no obvious reason identified to reject it. Other climatic data including temperature, relative humidity, and solar radiation were used as collected at the CGIS station, Butare, in the absence of similar nearby stations for comparison. Though the model used (HEC-HMS) does not allow entering soil texture/properties during the model set-up, the expected difference in response of different soils is addressed partially during calibration through adjustment of infiltration rates in different sub-catchments. Based on the findings of the data quality analysis, it was decided to limit the simulation work in the period between 1st August 2009 and 31st July 2010, with a condition of covering the entire calendar year. However, owing to lack of reliable long time observed flow data, the model validation could not be done in this study and all available data were used for model calibration. The model was validated using tracer based hydrograph separation results conducted by Munyaneza et al. (2012a), to compare the model results in terms of the runoff components (see Chap. 4).

6.2.2. Methods (HEC-HMS 3.5 and HEC-GeoHMS 5.0)

Two main tools were used in this study; the HEC-HMS 3.5 for the rainfall-runoff simulation and HEC-GeoHMS 5.0 for catchment delineation.

6.2.2.1 Hydrological model (HEC-HMS 3.5)

The latest available version HEC-HMS 3.5 was used in this study. Given rainfall values as input data, the HEC-HMS calculates outflow from the sub-catchment element by subtracting evaporation, calculating surface/direct runoff and adding baseflow. A full description of all components in HEC-HMS can be found in the user manual (USACE-HEC, 2010).

The Migina catchment was divided into 5 sub-catchments for computing evaporation and percolation, baseflow, transform and routing computation methods, and parameters were defined to convert rainfall into runoff. While running different scenarios, the HEC-HMS creates an output Data Storage System (DSS) file, which stores calculated data from all runs for a given project so that results from a preceding run can be directly compared with results from a new run. For purposes of reading and extracting the DSS file for results analysis, the HEC-DSSVue 2.0.1 tool was used.

The Hydrologic Engineering Center's Geospatial Hydrologic Modelling System (HEC-GeoHMS) Version 5.0 was used with ArcGIS 10.0 to derive river network of the catchment and to delineate sub-catchments of the Migina catchment. With GeoHMS, the project area was automatically delineated and its basin characteristics were generated (area, reach length, river slopes, etc). In addition, the HEC-GeoHMS created background map files and basin model files, which were later used by HEC-HMS to develop a hydrologic model. The sub-catchments delineation resulted into sub-catchments: Munyazi-Rwabuye (W380), Mukura (W410), Cyihene-Kansi (W400), Akagera (W650), and Migina (W640) (see Fig. 6.2).

6.2.3. Computation methods

To compute the different water balance components, the following computation methods, as referred to in the HEC-HMS literature, were applied to the sub-catchments (e.g. Yawson *et al.*, 2005) and reaches.

(*i*) *The Loss Method* (name as per HEC terminology as in the hydrological cycle a real loss does not exist) allows computing basin surface runoff, groundwater flow, total evaporation, as well as deep percolation out of the basin. The Soil Moisture Accounting (SMA) was selected as the appropriate approach to convert rainfall hyetograph into excess rainfall. In conjunction with the SMA, the canopy and surface losses (interception) were also considered and computed using simple canopy and simple surface methods (HEC, 2011).

(*ii*)*Transform Method* (runoff generation module) allows specifying how to convert excess rainfall into direct runoff. This method employs the Soil Conservation Service (SCS) technique (dimensionless unit hydrograph). The method requires only one parameter as input for each sub-catchment: lag time (T_{lag}) between rainfall and runoff in the sub-catchment (Eq. 6.1). The SCS developed a relationship between the time of concentration (T_c) and the lag time (T_{lag}). HEC-HMS includes an implementation of Snyder's Unit Hydrograph (UH). In his work, Snyder (1938) selected the lag, peak flow, and total time base as the critical characteristics of a UH. He defined a standard UH as one whose rainfall duration ($\frac{\Delta t}{2}$) is related to the basin lag (T_p) as shown in Equation 6.2.

$$T_{lag} = 0.6T_c \qquad\qquad (6.1)$$

$$T_p = \frac{\Delta t}{2} + T_{lag} \qquad\qquad\qquad (6.2)$$

where: T_{lag} = Lag time [min], T_c = Time of concentration [min], T_p = Basin lag [min], and $\frac{\Delta t}{2}$ = Duration of excess rainfall [min].

(iii) Baseflow method performs subsurface flow calculation. *The Linear reservoir baseflow method* was considered due to its simplicity and suitability for the SMA approach and was used to simulate continuously the recession of baseflow after a storm event.

(iv) The Muskingum-Cunge method, which is the routing technique used for the reaches, was selected in this model because of its numerical stability. The reach characteristics used were mainly produced by the HEC-GeoHMS (length and slope), and others borrowed from the previous publications carried out in the same catchment such as in SHER (2003), Van den Berg and Bolt (2010) and Munyaneza *et al.* (2010, 2011b & 2012a,b).

6.2.4 Basin model setup and simulations

6.2.4.1 Basin model

To run the HEC-HMS model, a basin file was created to specify the physical parameters of the catchment, and a map file to give the outline of the drainage areas was provided.

The Basin model supplies the physical datasets describing the physical properties of the watershed and typology of the stream network. In the present study, the basin model was created using the HEC-GeoHMS and then imported into the HEC-HMS with all its hydrologic elements: 5 sub-catchments, 10 junctions, 11 reaches, and a sink used to represent the outlet of a basin [node with inflow and without outflow] (Fig. 6.2). Where applicable, the junction elements were assigned to observed flow data, for use in comparison with simulated flows during the calibration process. In the model parameterization process, each hydrologic element was supplied with initial conditions and parameter values based on requirements of the different computation methods as discussed in the Section 6.2.3 above. Initial parameter values were selected based on the previous works where available, otherwise default values from the manual were applied. Table 6.1 shows the five computation methods: canopy, surface, loss, transform; and baseflow, the type of parameters used for each method, and values attributed to each parameter in the modeling process (initial and calibrated).

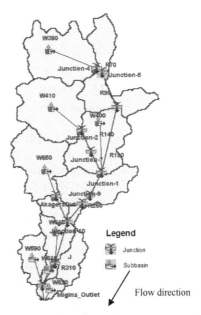

Figure 6.2 Migina catchment model set up in HEC-HMS.

6.2.4.2 Meteorological Model

The Meteorological Model was created after having created the Basin Model. The Meteorological model in HEC-HMS includes rainfall and evaporation methods to be used in the simulations (Arbind *et al.*, 2010).

In this study, the rainfall and evaporation data which are essential to simulate catchment processes were stored in the meteorological model. Twelve rain gauges and inverse distance method for rainfall computation were used in this model. The Priestley-Taylor method was used for computing total evaporation using temperature and radiation data. The current HEC-HMS 3.5 version allows total evaporation computation using temperature and radiation based method in combination with Soil Moisture Accounting (SMA) model.

6.2.5 Calibration methods

In the present study, a combination of manual and automated calibration techniques was used. Automated calibration known as "trial optimization" in HEC-HMS was used (Abushandi, 2011), to obtain optimum parameter values that give the best fit between observed and simulated flow volumes values (Ruelland *et al.*, 2008).

Given the availability of flow at the outlet of different sub-catchments, calibration was done catchment-wise starting from the farthermost upstream catchments (Munyazi, Mukura, and Akagera), since what happens upstream affects the results downstream. At the end of the calibration process, manually, the Nash–Sutcliffe Model Efficiency method *(NS)* was used to measure how the model fits the real hydrologic system (discussed under Sect. 6.2.5.1).

6.2.5.1. Model performance evaluation

The calibrated model performance was evaluated using the Nash–Sutcliffe Model Efficiency *(NS)* methods (Nash and Sutcliffe, 1970, Miao *et al.*, 2013). The *NS* is used to assess the predictive power of hydrological models. Mathematically, it is expressed as:

$$NS = 1 - \frac{\sum_{t=1}^{T} \left(Q_0^t - Q_m^t \right)^2}{\sum_{t=1}^{T} \left(Q_0^t - \overline{Q_0} \right)^2} \qquad (6.3)$$

where: Q_0^t is observed discharge at time t, $\overline{Q_0}$ is average observed discharge, and Q_m is modeled discharge at time t; all Q-variables have the unit runoff volume per time step (e.g. $m^3 s^{-1}$).

Nash–Sutcliffe efficiencies can range from $-\infty$ to 1. An efficiency of 1 ($NS = 1$) corresponds to a perfect match between the modeled and observed time series. Whereas, an efficiency of 0 ($NS = 0$) indicates that the model predictions are as accurate as the mean of the observed data. If the efficiency is less than zero ($NS < 0$) the observed mean is a better predictor than the model. More detailed information on NS can be found in Legates (1999), McCuen et al. (2006), Schaefli and Gupta (2007) and Kashid et al. (2010).

6.2.6 Tracer techniques for model validating

Hydrograph separations to separate the total runoff during floods in two or more components, based on the mass balances for tracer and water fluxes, were applied in Munyaneza et al. (2012a). Environmental isotopes (oxygen-18 (^{18}O) and deuterium (^{2}H)) and hydrochemical tracers (dissolved silica (SiO_2) and chloride (Cl^-)) were used as tracers.

The study showed that the results using the two-component hydrograph separations method using hydrochemical tracers are generally agree with the three-component separations using dissolved silica and deuterium. It was demonstrated that subsurface runoff is dominating streamflow generation during floods and baseflow periods. Particularly, more than 80% of the streamflow was generated by subsurface runoff (mainly shallow groundwater from valley floors) for two events that were investigated in detail (see Sect. 4.3.2). The tracer results were supported by shallow groundwater observations and the observed runoff coefficients. These results have been used to check the model simulation in this paper.

6.3. RESULTS AND DISCUSSIONS
6.3.1. Calibration Results

After running initial parameters over the simulation period and plotting the results against the observed flows, the first run did not yield acceptable results, and the initial parameters were subjected to calibration.

The model is calibrated using two years of rainfall and runoff data (August 2009 to June 2011). However, owing to the lack of reliable long-term flow observations, a classical model validation (e.g. split sample test) could not be done in this study and all available data were used for the model calibration. The suitability of the model was checked using the results of tracer investigations. This is not a strict model validation (like a split sample test as recommended by Klemes, 1986), however, it provided further insights into the model behavior and the model performance. The initial and finally calibrated parameters for each sub-catchment are presented in Table 6.1.

Table 6.1 Initial and finally calibrated parameter values for each sub-catchment.

Method	Parameter	Munyazi (W380)		Mukura (W410)		Cyihene-Kansi (W400)		Akagera (W650)		Migina outlet (W640)	
		Initial	Calibrated	Initial	Calibrated	Initial	Calibrated	Initial	Calibrated	Initial	Calibrated
Canopy	Max Storage (mm)	6	3	3	3	6	2	1	1	2	2
Surface	Max Storage (mm)	5	5	20	20	3	3	2	2	3	3
Loss	Soil (%)	60	35	60	35	60	35	60	35	60	55
	Groundwater 1 (%)	72	65	72	65	90	75	72	75	90	81.4
	Max Infiltration (mm/hr)	208	10	208	7.5	208	5.5	208	7.5	208	12
	Impervious (%)	0.5	3.5	0.5	2.75	0.5	6.3	0.5	8.5	0.5	4.5
	Soil initial Storage (%)	40	48	40	30	50	50	40	40	50	13.8
	Tension storage (mm)	22	15	22	5	8	5	22	4	18	5
	Soil Percolation (mm/hr)	2	4	2	2	1.75	0.8	2	1.75	10	1.97
	GW 1 Storage (mm)	307.5	237.0	307.5	50.0	307.5	150.0	307.5	100.0	307.5	303.6
	GW 1 Percolation (mm/hr)	3	2	3	3.6	0.04	0.5	0.7	1.3	0.3	8.159
	GW 1 Coefficient (hr)	150	4320	150	1296	150	1440	150	1014	150	1014
Transform	Lag Time [Min]	150	120	150	30	120.22	60	120	45	120.56	45
Baseflow	GW 1 Initial ($m^3 s^{-1}$)	0.002	0.004	0.028	0.021	0.358	0.782	0.002	0.204	0.273	0.373
	GW 1 Coefficient (hr)	8100	6480	8100	3240	2430	3746	1100	3240	8100	6480

Table 6.1 shows that despite the basin under consideration being a meso-catchment, the calibrated parameter values obtained varied from sub-catchment to sub-catchment, even for adjacent ones. The differences observed between the parameter values across the different sub-catchments were relatively small, except in some few cases where differences were considerable. The parameters with considerable differences include: (i) Maximum infiltration, (ii) Maximum soil storage (iii) GW1 storage, (iv) Lag-time, and (v) GW1 coefficient; and all the four formed sensitive parameters for the catchment. The initial values for soil moisture were collected from Mukura sub-catchment at Kadahokwa marshland. Because the soil parameters were collected in only one sub-catchment, we could not verify these parameter values for other sub-catchments, but had to rely on calibration.

Although correlation between infiltration rate and sub-catchment slopes was not strong ($r = 0.33$), the higher infiltration rate value is observed in the most lowland areas of the Migina sub-catchment, where the slopes are gentle and herbaceous and shrub crops dominate the land cover (almost 100%) (see Table 2.1).

Groundwater storage values were higher in sub-catchments that, due to their physiographic settings, have larger valley floors (Cyihene-Kansi and Migina). Sub-catchments of Mukura and Akagera showed small storage mainly due to their high surface runoff induced by very steep slopes. This translates also in their low contribution of the baseflow to the total flow.

The difference observed in the groundwater coefficients across the basin shows the varying behavior of the different sub-catchments in transforming groundwater into baseflow. The groundwater coefficient represents time lag applied on the linear reservoir for transforming water in groundwater storage into lateral flow, which generate baseflow in the river. The correlation analysis showed that a stronger correlation exists between the groundwater coefficient and the groundwater storage capacity ($r = 0.94$) compared with correlation between groundwater storage and size of the sub-catchment ($r = 0.39$).

With respect to Lag time, which represents the duration of time between the centroid of rainfall mass and the peak flow of the resulting hydrograph, it was noticed that despite a weak correlation between lag time and basin mean slope, the sub-catchment with very steep slopes (Mukura) showed faster response than those with gentle slopes (Munyazi).

6.3.1.1 Flow results

Generally, the model predicted well the flows volumes, though difficulties in matching simulated and observed daily flows were observed.

Particular attention was given mainly to control points that collect water from more than one sub-catchment (Cyihene-Kansi and Migina outlets). During the calibration process, we tried to minimize the absolute values of the residuals of the observed flow volumes. In addition, the *NS* was used to better evaluate the performance of the calibrated model. Table 6.2 summarizes the obtained *NS* coefficients and total flow residual values for each discharge computation point in the basin.

Table 6.2 Residual values for each discharge computation point with corresponding *NS*. The simulation period is 12 months (1ˢᵗ August 2009 to 31ˢᵗ July 2010). The positive sign (+) means that the model overestimated the flows while the negative sign (-) means that the model underestimated the flows.

Sub-catchment name (code)	Station name	Total observed Q (mm a⁻¹)*	Total simulated Q (mm a⁻¹)*	Residual in % of total observed Q	NS [-]
Munyazi (W380)	Rwabuye	64.98	67.11	3.28	0.38
Mukura (W410)	Mukura	60.32	59.20	-1.86	0.62
Cyihene-Kansi (W400)	Kansi	366.93	382.63	4.28	0.51
Akagera (W650)	Akagera	296.89	322.35	8.58	0.61
Migina outlet (W640)	Migina	324.71	318.98	-1.76	0.65

The discharges are expressed in mm per entire simulation time.

Table 6.2 shows that the model performed reasonably well in simulating total flow volumes (Roy *et al.*, 2013). The residues in % of total observed range between -1.86% and 8.58% of observed flow. Results indicated by *NS* coefficients also depicted reasonable model performance in most cases (*NS* > 0.5) with the exception of Munyazi sub-catchment (*NS* = 0.38). Furthermore, the model simulated well the baseflow while reproducing at the same time the observed peaks in term of timing and quantity (Fig. 6.3). For instance, the model was able to reproduce the peak recorded at all stations on 2ⁿᵈ May 2010 as shown in Fig. 6.3. Similar results were obtained by Munyaneza *et al.* (2012a), who investigated the peaks discharge in the same catchment and observed the same peaks at the same time as in the current study (see Sect. 4.3.1).

In individual sub-catchments, the model performed relatively well in sub-catchments Akagera, Mukura and Migina (the outlet) with *NS* coefficients of 0.61, 0.62 and 0.65, respectively.

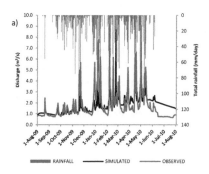

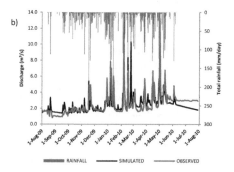

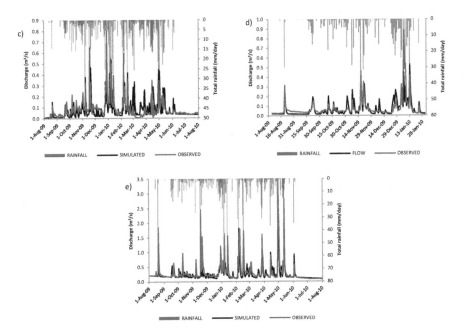

Figure 6.3 The simulated and observed hydrographs at a) Cyihene-Kansi, b) Migina outlet, c) Munyazi, d) Mukura, and e) Akagera sub-catchments.

Moreover, baseflows were also well simulated in most cases, with the exception at Cyihene-Kansi (Fig. 6.3a) and Migina outlet (Fig. 6.3b) where the model overestimated and underestimated the baseflow in dry seasons (June-July 2010), respectively. The main reason that our model simulates high and low recession of baseflow (at Cyihene-Kansi and Migina outlet) after a storm event may be linked to the inflexibility of the model structure. The results might have been improved by using a flexible model structure, e.g. FLEX-Topo (Savenije, 2010; Fenicia *et al.*, 2008a,b and 2010; Gharari *et al.*, 2013; Gao *et al.*, 2013). Savenije (2010) demonstrated that FLEX-Topo model allows the groundwater time scales to be lumped and determined by manual calibration on the recession curve.

6.3.2 Simulated water budget components
Recalling one of the main objective of water resources assessment (determination of water resources availability at local/sub-catchment level), the catchment water budget components from the model results were analyzed. The components are the total rainfall, actual evaporation and percolation, direct runoff, baseflow, and total flow. The quantities are presented in Table 6.3 and represent the total volume over the simulation period of 12 months (1st August 2009 to 31st July 2010).

Table 6.3 Budget components quantities for all sub-catchments in the simulated period of 12 months.

Sub-catchment name (code)	Total Rainfall (mm a⁻¹)	Evaporation + Percolation (mm a⁻¹)	Direct Runoff (mm a⁻¹)	Base flow (mm a⁻¹)	Total Flow (mm a⁻¹)	Base flow in % of the Total Flow	Direct flow in % of the Total Flow	Correlation Rainfall-runoff (-)
Munyazi (W380)	1453.0	1408.1	44.9	19.7	64.6	30.4	69.5	0.94
Mukura (W410)	1665.5	1622.5	43.0	16.2	59.2	27.4	72.6	0.96
Cyihene-Kansi (W400)	1456.6	1309.7	146.9	267.6	414.4	64.6	35.4	0.73
Akagera (W650)	1507.0	1382.1	125.0	127.5	252.5	50.5	49.5	0.97
Migina outlet (W640)	1415.2	1353.8	61.5	138.1	199.6	69.2	30.8	0.96

Table 6.3 shows that contributions of direct runoff and base flows vary from sub-catchment to sub-catchment, despite the small size and closeness of the sub-catchments. Table 6.3 shows that the outflows for Mukura and Munyazi sub-catchments depend highly on direct flow, whereas baseflow contribution was evaluated only at 27.4 and 30.4% of total flow, respectively. The observed dominance of high direct runoff in both sub-catchments may be attributed to the urbanization observed in the catchment areas such as Ngoma, Matyazo and Rwabuye towns (Fig. 2.3 and Table 2.1), resulting in relatively large areas of mainly imperviousness surfaces for rural catchments of 2.8% for Mukura and 3.5% for Munyazi of the total catchment areas. Opposite results were observed at Cyihene-Kansi and Migina outlet sub-catchments where the baseflow contributes 64.6 and 69.2% of total outflow, respectively (see Table 6.3 and Fig. 6.4).

In the absence of enough data to validate the model, an attempt was made to compare outputs of the present study with those obtained using other techniques than computational modelling. Two rainfall events were investigated during the rainy season in 2010 and 2011, using flow data collected at Kansi and Migina flow stations. The results showed that direct runoff component did not exceed 33.7 and 28.7% of the total event runoff, respectively. The model estimations of 35 and 31%, respectively, are close to the values obtained by tracer methods (Fig. 6.4). These values are the %-values for exactly these two events and not for the longer simulation period.

Note that in the HEC-HMS model output, the runoff components use the terms direct runoff and baseflow (Merz et al., 2009), but this is not in line with the terminology used in tracer based analysis (e.g. Munyaneza et al., 2012a) in which the components were defined in a process-oriented way (subsurface runoff, later flows, etc.). In presenting the comparison here, we have chosen to follow the terminology as used in HEC-HMS. Even though the results were slightly different, in the two sub-catchments (Cyihene-Kansi and Migina) as shown in Table 6.3, tracer methods confirmed the dominance of baseflow (HEC-HMS terminology) contribution to total streamflow (see also Sect. 4.3.2). This was due to more groundwater contributions to those two downstream sub-catchments in contrast to the upstream located sub-catchments. In addition, the convergence of modelling and tracer techniques shows that tracer data can serve as multi-response data to assess and validate a model, which was also concluded by Uhlenbrook and Leibundgut (2002) and Uhlenbrook et al. (2004). Hence, the model can be trusted from a process point of view and, therefore, seems useful for water resources planning purposes in the Migina catchment. The high contributions of baseflow to total flow translate into high reliability/security of water resources even during dry seasons, hence explaining the predominance of agricultural activities (91.2%) in the two sub-catchments (Cyihene-Kansi and Migina) as also found by Munyaneza et al. (2011b). This high contribution of baseflow to total flow also confirms the perennial river system observed in the Migina catchment during the study period, which is also supported by Munyaneza et al. (2012a).

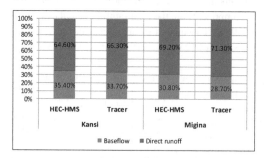

Figure 6.4 Comparison of flow components results using HEC-HMS model (current study) and hydrochemical tracer method (obtained from Munyaneza et al., 2012a, see Chap. 4 of this thesis) for two investigated events in the rainy season in 2010 and 2011, using flow data collected at Kansi and Migina flow stations.

Looking at other parts of the basin, for the Akagera sub-catchment (32.2 km^2), the baseflow and direct flow contribute about equal amounts to the sub-catchment outflow (50.5% and 49.5%, respectively). Compared to other sub-catchments within almost the same size, Munyazi (38.6 km^2) and Mukura (41.6 km^2), Akagera (32.2 km^2) has a considerable high direct runoff (3 times the direct runoff of the other two) mainly attributed to the steep slopes (20.8%) and to the high portion of impervious (8.5%) areas in this sub-catchment (see Table 2.1). However, nothing fully explains the higher baseflow contribution to the total runoff compared with Munyazi and Mukura sub-catchments, apart from the three sub-catchments are different in nature (e.g. topography, shape of river channel).

Cyihene-Kansi sub-catchment (69.6 km^2) yields a lot of water compared with the other 4 sub-catchments. Its high outflow of 414.4 mm over the simulation period is explained by its high amount of baseflow (267.6 mm), and higher direct flows (146.9 mm) resulting most probably from its bigger size than other sub-catchments (Tables 2.1 and 6.4).

In general, the Akagera sub-catchment simulations gave better results with high correlation between rainfall-runoff (r= 0.97) than the other four sub-catchments (Munyazi, Cyihene-Kansi, Mukura and Migina) (see Table 6.3). The better result in this sub-catchment may be partly attributed to the Akagera river channel of a rectangular shape that favors more accurate discharge measurements compared with other rivers in the catchment. The other reason could be that the used daily time step is less suitable for small steep catchments.

6.4. CONCLUDING REMARKS

In this study, the model HEC-HMS version 3.5 hydrologic modelling software was applied to the Migina meso-scale catchment, and the model parameters for total evaporation (Soil Moisture Accounting method) and baseflow (linear reservoir) were calibrated using the observed stream flows. The model performed reasonably well over the calibration period by reproducing the observed flow volumes and simulating the observed peaks in terms of timing and quantity. The HEC-HMS model was applied to 5 sub-catchments and the model results were compared with tracer results in two sub-catchments (Cyihene-Kansi and Migina), however, the model was not validated in a classical way due to the lack of reliable data (cf. Du *et al.*, 2007), but checked using the results of tracer investigations for its plausibility. This is not a strict model validation (like a split sample test as recommended by Klemes, 1986), however, it provided further insights into the model behavior and the model performance. Based on the performance of the HEC-HMS model and the tracer method comparison, the present study concluded that the framework works effectively well in the meso-scale Migina catchment, but needs to be flexible in its structure for simulating the recession and baseflow. The conclusion was that flexible models should probably work better in meso-scale catchments like Migina than models which are not flexible (having fixed applications). This was supported by Fenicia *et al.* (2011) who proposed flexible models because they allow the hydrologist to hypothesize, build and test different model structures using combinations of generic components. They said that flexible models are particularly useful for conceptual modelling at the catchment scale, where limitations in process understanding and data availability remain major research and operational challenges (also in line with Uhlenbrook *et al.*, 1999).

The simulation results gave indication of zones of high surface runoff and for recharge/baseflow generating areas. Those zones present potential areas where catchment protection interventions can be implemented. For example, interventions leading to protection of the water sources can be implemented in the zones of recharge where infiltration, recharge and temporary groundwater storage are higher. Areas of higher direct runoff, mainly due to the slopes, may also be suitable for interventions leading the reduction of slopes by terracing, and hence increasing infiltration and subsequent recharge.

Moreover, at the meso-scale catchment level, considerable disparities in the parameters and hydrological processes exist. Lumping the entire Migina catchment would lead to missing important aspects of some of the sub-catchments and, subsequently, potentially misinforming the planning and decision making processes. Depending on the purpose of the assessment and the intended use of the information to be generated, individual units at an appropriate scale may require particular attentions even in very small catchments.

Continuous quality assurance and control of hydrological and weather data sets recorded at different stations in the entire catchment is of great importance for the future.

Chapter 7

CONCLUSIONS AND RECOMMENDATIONS

7.1. SUMMARY OF THE MAIN CONCLUSIONS

The aim of this research is to explore hydrological trends and climate linkages for catchments in Rwanda, with a particular focus on understanding dominant hydrological processes in the meso-scale Migina catchment, Southern Rwanda. The research was focusing on understanding the hydrological processes within a meso-scale catchment for sustainable water resources planning and management by using experimental research methods. Specifically, the study emphasizes on the investigation of the relationship between trends in hydrologic variables and climate variability, quantifying the runoff components and identifying the dominant hydrological processes in a meso-scale catchment. The water resources availability in the meso-scale Migina catchment was also assessed in this research using catchment modelling. Understanding the hydrological processes in an ungauged meso-scale catchment requires a multi-method approach, which is described in Chapters 4-6.

This study identifies the linkages between climatic data and streamflow data over the main Rwandan catchments (Chap. 3). The used Mann-Kendall tests to detect trends in both climate and streamflow demonstrate the importance of the different periods of the available time series. In general, trends in streamflow are stronger for the longer period of data (e.g. for 1961-2000 compared to 1971-2000). The results show strong increasing trends in temperature (1971 to 2006) and both increasing and decreasing trends (e.g. Pt2 and Pt4 in November) in rainfall (1961 to 2008) as demonstrated in Table 3.5.

In the all stations, there is both increasing and decreasing trends and shift change points in many hydrological variables for the investigated catchments during the period 1980–1990 as shown in Table 3.4. The observed change points of most stations (Table 3.4) are likely due to human activities such as increased agricultural activities, irrigation, or water supply, but cannot be attributed to climatic changes (i.e. rainfall) (Zhang et al., 2011). It could also be due to the fact that the observed correlation within the data was found to be statistically significant at a 5% confidence level for many stations. Streamflow was found to be positively correlated with the total rainfall and negatively correlated with the mean temperature due to the similar variation observed between them as also confirmed by Mutabazi et al. (2004). However, the observed similarity in streamflow and rainfall variation implies that changes in streamflow should be linked with corresponding changes in the rainfall conditions. This gives a hint to the sensitivity of Rwanda water resources for future climatic conditions. This was supported by Pettitt tests, which reveal that abrupt change points of most stations occurred in the recent 1980s–1990s (Table 3.4), which is related to the period of data gaps (Table 3.1) and a period of intensive human development activities (PRSP, 2007). The results revealed that some climate variables could not easily explain the observed trend behavior (see Table 3.5). However, the existing gauging stations are providing unreliable data as rating curves were not updated (some of them are 30 years old and more). No regular discharge measurements are taken to update existing rating curves referring to Equation 2.1. The selected 4 stations for data analysis had modified rating curves which were updated and corrected. Some modified rating curves were obtained for instance at St3 (RIWSP, 2012c), St5, St6 and St8 (RIWSP, 2012a). Therefore, we conclude that for stations St1, St2, St4, and St7 it is not possible to correct and adjust the raw data in such a way that further analysis regarding trend analysis could be carried out. New data with good quality are also needed for future better water resources management in Rwanda. Hence, the country needs to be much more sensitive about data collection.

Tracer methods were applied in conjunction with hydrometric measurements for identifying dominant runoff generation processes in the meso-scale Migina catchment (Chap. 4). The two- and three-components hydrograph separation models using hydrochemical (dissolved silica and chloride) and isotope (deuterium and oxygen-18) tracers show that intensive water sampling (hourly) during events is essential. The results of this study demonstrated the importance of subsurface flows for streamflow generation in the study area. It also shows the value of detailed hydrological data collection over two whole rainy seasons using different tracers and hydrometric observation to understand dominant hydrological processes. Furthermore, it demonstrated the significance of considering spatial and temporal variations of rainfall in the hydrograph separations (Figures 4.7 and 4.8); this is of greater importance in meso-scale catchments than in small headwaters. Oxygen-18 (^{18}O) and deuterium (^{2}H) were found to be suitable tracers to detect event vs. pre-event water sources. Additionally, it was found that groundwater has two different origins: one source originates from a near stream location in the valleys (shallow groundwater) and the other source is deep groundwater sampled at piezometers and springs located towards the upper part of the hillslopes (Sect. 4.3.3). The significant groundwater recharge during the wet seasons led to the perennial river system observed in the catchment. The isotope analysis showed that all runoff components including baseflow are dependent on wet season rainfall.

Furthermore, this research shows that the Migina catchment peak runoff discharge can be predicted with a suggested simple empirical approach (based rational method) and by using data from field investigations (Chap. 5). Land cover and hydrological soil groups analyses in the Migina catchment show that the Migina catchment is dominated by agriculture activities (92.5%) while forest occupy 5%, grass/lawn 2% and buildings cover 0.5% only. The runoff coefficient obtained after an overlay of a land cover map and a soil type map was estimated to 0.25. This value falls in the range for an agricultural dominated catchment as found by Purdue *et al.* (2007). The river peak discharge generated in the Migina catchment on average every 10 years is around 16 m^3 s^{-1} (62.6 l/km^2 s) by use of an area reduction factor and the rainfall intensity observed in the Migina catchment. Results found by Munyaneza *et al.* (2012a) showed that the maximum discharge in the Migina catchment was around 15 m^3 s^{-1} using discharge data measured from 1st August 2009 to 31st June 2011. The time of concentration was estimated to 3.5 hours. It is recommended that this approach is applied to similar agricultural dominated catchments in the region and results could be used for flood management and support decision making. This recommendation is based on rational method plausibility check which has been performed using field observations (Verdoodt and van Ranst, 2003, Nahayo, 2008, Nahayo *et al.*, 2010, Munyaneza *et al.*, 2012a).

For further assessing runoff and water resources availability on each sub-catchment level, we recommended developing river discharge models using simple methods such as conceptual hydrological modelling to simulate continuously the rainfall runoff of this catchment. Based on this recommendation, the process-based semi-distributed hydrological (HEC-HMS 3.5) model was developed in this research (Chap. 6). The model is expected to assist as a tool for water resources planning and decision making processes in this catchment.

The model HEC-HMS version 3.5 calibrated the parameters for total evaporation (Soil Moisture Accounting method) and baseflow (linear reservoir) using the observed stream flows. The model performed reasonably well over the calibration period by reproducing the observed flow volumes and simulating the observed peaks in terms of timing and quantity. The HEC-HMS model was applied to 5 sub-catchments and the model results were compared with tracer results in two sub-catchments (Cyihene-Kansi and Migina), however, the model was not validated in classical way due to the lack of reliable data (cf. Du *et al.*, 2007). Based on the performance of the HEC-HMS model and tracer method comparison, the present study concluded that the framework works effectively well in the meso-scale Migina catchment but needs to be flexible in its structure for simulating the recession baseflow.

The simulation results gave indication of zones of high surface runoff and for recharge/baseflow generating areas. Those zones present potential areas where watershed protection interventions can be implemented. For example, interventions leading to protection of the water sources can be implemented in the zones of recharge where infiltration, recharge and temporary groundwater storage are higher. Areas of higher direct runoff, mainly due to the slopes, may also be suitable for interventions leading the reduction of slopes by terracing, and hence increasing infiltration and subsequent recharge.

Moreover, at the meso-scale catchment level, considerable disparities in the parameters and hydrological processes exist. Lumping the entire Migina catchment would lead to missing important aspects of some of the sub-catchments and, subsequently, potentially misinforming the planning and decision making processes. Depending on the purpose of the assessment and the intended use of the information to be generated, individual units at an appropriate scale may require particular attentions even in very small catchments.Continuous quality assurance and control of hydrological and weather data sets recorded at different stations in the entire catchment is of great importance for the future.

7.2. RECOMMENDATIONS

The conclusions above lead to the following recommendations:

- Researchers should include the Congo basin (20% on Rwandan territory) for studying the hydrological trends and climate linkages for meso-scale catchments in Rwanda. We looked at the Nile basin part only. This would help to understand the influence of El Niño and La Niña to Rwandan climate and streamflow variability.
- Due to data scarcity, lack of quality and missing data observed in the analysis, (1) the government of Rwanda (GoR) should ensure that hydro-meteorological data are collected according to the international standards; (2) the Rwanda Meteorological Agency (RMA) should make sure that all meteo data recorded by private institutions (often by individuals) are also kept in their database if quality controlled; (3) the GoR should motivate academic institutions to introduce professional and technical training in the field of hydrology, meteorology, and other water related issues, and (4) private sector organizations and NGOs should invest in these training as there are required around the country for future sustainable water resources management and planning.
- More studies on runoff generation processes are needed in the micro-scale and meso-scale catchments in the different regions of Rwanda (more Migina-type catchments). Hence, intensive water sampling (hourly) during events is essential. The whole rising limb, peak and recession limb need to be captured completely for the event in order to gain better understanding of runoff generation processes. However, different geographic sources of runoff need to be observed before, during and after the events.
- Researchers should test the proposed rainfall-runoff model, which is a semi-distributed hydrological (HEC-HMS 3.5) model, in various case studies.
- Further research should be done in other catchments in Rwanda rather than in agriculture catchments as tested in Chapter 5 of this research; and finally
- Decision makers should use the findings from this study for future sustainable development of water resources policies/strategies in Rwanda. This will contribute to the mission of the Rwanda Ministry of Agriculture (MINAGRI) in their objective of land consolidation and the Rwanda Vision 2020, which is committed to reduce dependency on agriculture through diversification of productive sectors and employment diversification. This will reduce the pressure on land and water resources, given that agriculture accounts for about 70% of the total water use at national level.

SAMENVATTING

Ruimte en temporele variatie van hydrologische processen en watervoorraden in Rwanda, aandacht voor het Migina stroomgebied

Snelle bevolkingsgroei in Rwanda vergroot de vraag naar water voor huishoudelijke, agrarische en industriële gebruik en veroorzaakt waterschaarste. Er is een toenemende druk op alle natuurlijke hulpbronnen, waaronder water. Voor een goede schatting en beheer van watervoorraden in stroomgebieden, is het essentieel om belangrijke hydrologische processen te identificeren. Daarnaast is het ook heel belangrijk om de huidige en toekomstige verdeling van het water te begrijpen. In Rwanda is er een gebrek aan ruimtelijke en temporele hydrologische en klimatologische gegevens, vooral na de genocide in 1994, wat studies op dit gebied belemmerd.

Het doel van dit onderzoek is om de hydrologische trends en klimaatverbanden voor de belangrijkste stroomgebieden in Rwanda te verkennen, en een bijdrage te leveren aan het begrip van dominante hydrologische processen in de meso-schaal Migina stroomgebied, Zuid-Rwanda. De studie legt nadruk op het onderzoek naar de relatie tussen de hydrologische variabelen en klimaatvariabiliteit, het kwantificeren van de afvoerkomponenten en het identificeren van de dominante hydrologische processen in een meso-schaal stroomgebied. De beschikbaarheid van watervoorraden in het mesoschaal Migina stroomgebied is vastgesteld met behulp van stroomgebiedsmodellen.

Een "multi-method" benadering is gevolgd om de onderzoeksdoelstellingen te bereiken, gebruikmakend van experimentele en modelleringstechnieken,. In de periode van april 2009 tot juli 2009, zijn er verschillende hydrologische instrumenten in het Migina stroomgebied geïnstalleerd, waaronder doorlopende rivierafvoermetingen. In het kader van dit onderzoek is een studie naar de relatie tussen de rivierafvoer en het klimaat uitgevoerd, gebruikmakend van langdurige rivierafvoermeetingen met een minimum van 30 jaar (1961-2000) voor geheel Rwanda (26.338 km^2). De Mann-Kendall (MK)-test is uitgevoerd om trends in de tijdsreeksen te ontdekken en the Pettitt test is gedaan om de timing van de veranderingspunten te identificeren. De relaties tussen elk van de klimatologische en hydrologische variabelen is onderzocht met behulp van Pearson correlatie. De analyses tonen significante trends voor de klimatologische en hydrologische variabelen en een stijgende trend van rivierafvoeren. De rivierafvoer blijkt positief gecorreleerd met de totale neerslag en negatief gecorreleerd met de gemiddelde temperatuur. Pettitt tests tonen aan dat abrupte veranderingspunten van de meeste stations zich voor deden in de jaren 80 en 90, die gerelateerd zijn aan een periode van intensieve menselijke activiteit in Rwanda.

Met behulp hydrometrische data en moderne tracer methoden, worden de hoeveelheden van elke afvoerkomponent bepaald en de dominante hydrologische processen in de Migina meso-schaal stroomgebied geidentificeerd. De tracer methode maakt gebruik van de massabalansvergelijking en eindonderdeel meng-analysen voor twee-en drie-componenten hydrograafscheidingsmodellen (gebruikmakend van deuterium (^{2}H), zuurstof-18 (^{18}O) gesuspendeerd (Cl$^-$) en opgelost silica (SiO$_2$)). De resultaten tonen aan dat ondergrondse afvoer dominant is in de totale afvoer, zelfs tijdens overstromingen. Tijdens twee onderzochte vloedgolven (1 tot 2 mei 2010 benedenstrooms van het Cyihene-Kansi sub-stroomgebied en 29 april - 6 mei 2011 benedenstrooms van het Migina stroomgebied), blijkt dat meer dan 80% van de afvoer is gegenereerd door ondergrondse afvoer. Deze dominantie van de ondergrondse afvoer is in overeenstemming met de waargenomen lage waarden voor de afvoer coëfficiënt voor beide vloedgolven (16,7 en 44,5%). Grondwateraanvulling vind daarom vooral plaats tijdens het natte seizoen en leidt tot een pereniaale rivierafvoer in het Migina stroomgebied.

In hetzelfde Migina stroomgebied is een eenvoudige rationele methode met ruimtelijke correctie gebruikt om de rivier piekafvoer te voorspellen. Landbouw is, met ongeveer 92,5% het dominante landgebruik in het stroomgebied. De resultaten tonen een gewogen afvoer coëfficiënt van 0,25, de tijd van de concentratie 3 uur en 26 minuten en een piekafvoer van 16,12 m^3 s^{-1} (62,6 l/km^2 s). Voor het bepalen van afvoer en waterbeschikbaarheid per sub-stroomgebied, hebben we gebruik gemaakt van het afvoermodel Hydrologisch Engineering Center-het Hydrologisch Modelling System (HEC-HMS). Dit model kan helpen in besluitvorming processes in het waterbeheer van dit stroomgebied. Het HEC-RAS model was geselecteerd omdat het de capaciteit heeft om de ruimtelijke variatie van afvoer genererende processen in kaart te brengen , het eenvoudig in opzet, heeft weinig gegevens nodig, en het is gratis software pakket.

Na het definieren van de parameters en initiele condities van elke sub-stroomgebied, zijn de dagelijkse afvoeren gesimuleerd met de Soil Moisture Accounting methode en een lineair reservoir model. Het HEC-HMS version 3.5 model maakt gebruik van de Soil Moisture Accounting methode die de de verdamping en infiltratie berekent, het lineaire reservoir model berekent de basisafvoer en de Muskingum-Cunge routine is gebruikt voor het routen van de afvoer. We hebben gebruik gemaakt van gegevens van 12 regen-, en 5 afvoerstations, die zijn verzameld gedurende een periode van twee jaar (mei 2009 tot juni 2011). Het stroomgebied was opgedeeld in vijf sub-stroomgebieden afhankelijk van de lokatie van de vijf afvoerstations.

De modelparameters zijn gekalibreerd met behulp van waargenomen afvoermetingen voor elk sub-stroomgebied, en de resulterende afvoeren werden vergeleken met de waargenomen afvoeren. De resultaten verkregen door de calibratie werden acceptabel gevonden voor vier afvoerstations met een Nash-Sutcliffe Model Efficiency (NS) van > 0,65. Wegens het ontbreken van voldoende en betrouwbare gegevens voor langere periodes, werd een model validatie (split-sample test) niet uitgevoerd. In plaats daarvan, zijn de resultaten vergeleken met de resultaten van de hydrograafscheiding met behulp van tracers. Het model werkt redelijk goed voor de kalibratieperiode, zowel in het simuleren van de basis afvoer, reproduceren van de waargenomen afvoervolumes, en simuleren van de waargenomen pieken qua timing en hoeveelheid. Daarnaast werden aanzienlijke verschillen in de parameters en hydrologische processen waargenomen over de vijf sub-stroomgebieden, ondanks dat de sub-stroomgebieden relatief klein zijn. Dergelijke verschillen rechtvaardigen de noodzaak om stroomgebieden onder te verdelen op het laagst mogelijke niveau, indien deze parameters en hydrologische processen de basis vormen voor de besluitvorming in het water resources planning en management in het stroomgebied.

De kennis die met deze studie is gegenereerd, is voor beleidsmakers van essentieel belang voor het bepalen van nationale waterbeleid en -strategieën voor een beter waterbeheersplanning en -beheer. De opgedane kennis in deze studie zal zoveel mogelijk worden overgedragen naar andere Rwandese stroomgebieden om zo bij te dragen aan de 2020 visie van Rwanda en de Millennium Development Goals (MDG's). Het Rwanda Vision 2020 streeft naar verminderen van de afhankelijkheid van de landbouw door middel van diversificatie van de werkgelegenheid. Dit zal hopelijk de druk verminderen op de watervoorraden, gezien het feit dat de landbouw goed is voor meer dan 70% van het totale watergebruik op nationaal niveau.

Omar Munyaneza
UNESCO-IHE, Delft, The Netherlands
April 2014

....

The summary in Dutch was translated by Dr. Marloes Mul, Delft, The Netherlands

INCAMAKE (SUMMARY IN KINYARWANDA)

Ihindagurika mu gihe n'intera ry'uruhererekane rw'inzira z'amasoko y'amazi n'umutungo kamere w'amazi mu Rwanda, hibandwa cyane cyane ku kibaya cy'igishanga cya Migina

Ubwiyongere bwihuse bw'abaturage mu Rwanda burongera ikenerwa ry'amazi mu mikoreshereze yo mu ngo, mu buhinzi no mu nganda, bukanatuma amazi aba make (adahaza). Kuri ubu bwiyongere bukabije hiyongeraho n'ikoreshwa ridasiba ry'umutungo kamere harimo n'amazi. Kugaragaza uruhurirane n'imikoranire y'uruhererekane rw'inzira z'amasoko y'amazi ni ingirakamaro kugira ngo habeho imicungire iboneye n'isuzumagaciro ry'umutungo kamere w'amazi yo mu masoko. Ni ngombwa kandi gusobanukirwa uko amazi akwirakwizwa mu bibaya by'inzuzi ukurikije igihe cyangwa intera kuri ubu no mu gihe kizaza. Kutagira imibare y'ibipimo irebana n'iby'amasoko y'amazi n'ibihe mu Rwanda, ukurikije uko imyaka yagiye ikurikirana n'intera, by'umwihariko mugihe na nyuma ya jenoside yo muri 1994, bidindiza bikomeye inyigo zikorwa kubijyanye n'ubumenyi bw'amasoko y'amazi muri aka gace.

Intego y'ubu bushakashatsi ni ugusesengura (gusuzuma) imiterere y'umutungo w'amazi n'isano ifitanye n'ibihe ku bibaya (ibishanga) byo mu Rwanda (Rwandan catchments), no gutanga umusanzu mugufasha gusobanukirwa imiterere n'imikoranire yiganje mu uruhererekane rw'inzira z'amasoko y'amazi ziri mu gishanga cya Migina, gihereye mu majyepfo y'u Rwanda. By'umwihariko, ubu bushakashatsi bugamije gusesengura isano iri hagati y'imihindagurikire y'umutungo w'amazi n'imihindagurikire y'ibihe, kugaragaza ingano y'ibice byuko amazi atemba mugihe cy'imvura (runoff components) no kugaragaza inzira ziganje muri urwo ruhererekane rw'inzira z'amasoko y'amazi mu gishanga kiringaniye (meso-scale catchment). Muri ubu bushakashatsi kandi hasuzumwe byumwihariko umutungo kamere w'amazi uboneka mu kibaya cy'igishanga kiringaniye cya Migina hifashishijwe ikoranabuhanga rigezweho (catchment modelling).

Kugira ngo intego z'ubu bushakashatsi zigerweho, twifashishije urukomatanye rw'uburyo bunyuranye bukoreshwa mu bushakashatsi (multi-methods) harimo ibikorwa by'igerageza ryakorewe aho igishanga kiri (experimental) ndetse n'ikoranabuhanga (modelling activities). Ibikoresho bitandukanye bijyanye n'ubumenyi bw'umutungo w'amazi (hydro-meteorological instrumentations) byashyizwe ku kibaya cy'igishanga cya Migina hagati ya Mata na Nyakanga 2009. Imibare y'ibipimo by'imiterere y'ibihe ndetse n'ishoka ry'umugezi byarafashwe kandi biracyakomeza. Muri ubu bushakashatsi, inyigo ku byerekezo by'itemba ry'imigezi no ku isano bifitanye n'ibihe mu Rwanda hose (26,338 km^2) yarakozwe, hifashishijwe ahari hubatse ibipimo (amasitasiyo) bipima ingano y'imvura niy'amazi atemba mu migezi. Ariko hatoranijwe gusa amasitasiyo afite imibare y'ibipimo byo mu gihe kirekire kitari munsi y'imyaka 30 (byafashwe hagati ya 1961 na 2000).

Imihindagurikire n'igihe cy'ahabereye impinduka kubihe byo mu Rwanda byakorewe ubushakashatsi hifashishijwe igerageza rya Mann-Kendall n'irya Pettit (Mann-Kendall and Pettitt tests). Amasano ari hagati ya buri bihe n'ibice bigize ubumenyi bw'amasoko y'amazi yarasuzumwe hifashishijwe igereranya ry'umushakashatsi Pearson (Pearson correlation). Ibyavuye muri ubu bushakashatsi byagaragaje imihindagurikire ikabije (significant trends) ku bihe byo mu Rwanda no mu mibare irebana n'amazi. Muri rusange kwiyongera gukabije kw'imihindagurikire y'ishoka (itemba) ry'imigezi kwagaragaye iyo ufashe imibare y'ibipimo byigihe kirekire (1961-2000), naho kugabanyuka kwagaragaye hakurikijwe imibare y'ibipimo byafashwe mu gihe gito (1971-2000). Igerageza rya Pettit (Pettit tests) ryo ryagaragaje ihinduka ritunguranye (abrupt change points) mu bihe byo mu Rwanda. Ibi bikaba bigaragarira ku masitasiyo menshi ari mu gihugu aho yerekana ko impinduka ikabije yabaye cyane mu myaka ya za 80 na 1990. Twanzuye ko iri hinduka rishobora kuba rifitanye isano n'igihe cyo kwiyongera kw'ibikorwa bya muntu mu Rwanda nk'iterambere mu buhinzi (ikoreshwa ry'amazi mu kuhira imyaka) n'iterambere mu iyubakwa ry'imijyi.

Ubu bushakashatsi nanone bwagerageje gutanga imibare y'ibipimo by'ibice bigaragara mu itemba ry'amazi y'imvura igihe ageze kubutaka (runoff components), no kugaragaza igice kiyatwara kurusha ibindi mu uruhererekane rw'inzira z'amasoko y'amazi mu kibaya cy'igishanga cya Migina. Hakoreshejwe imibare y'ibipimo byafashwe ndetse n'uburyo bugezweho bwa taraseri (hydrometric data and modern tracer methods). Ku bijyanye no gukoresha uburyo bugezweho (tracer methods), twakoresheje ubuhanga bwo kugabanyamo ibice bibiri na bitatu byuko amazi yiyongera mu mugezi igihe imvura igwa (two- and three-component hydrograph separation models) hifashishijwe ibipimo by'ubutabire (deuterium (^{2}H), oxygen-18 (^{18}O), chloride (Cl$^-$) and dissolved silica (SiO$_2$)).

Ibyavuye muri ubu bushakashatsi bigaragaza ko amazi yinjira mu migezi anyuze mubutaka aruta amazi yose yatembye kubutaka yinjira mu migezi. Ibi bikaba byaranagaragaye no mu gihe cy'imyuzure (imvura nyinshi). Hagaragaye ko hejuru ya 80% by'amazi atemba mu migezi aba yatembye abanje kwinjirira mu butaka. Ibi bikaba byaragaragaye hifashishijwe ibipimo (imibare) byafashwe mugihe cy'imyuzure ibiri yabaye mugihe ubu bushakashatsi bwakorwaga (ku itariki ya 1 n'iya 2 Gicurasi 2010 ku kibaya cy'igishanga cya Cyihene i Kansi no kuya 29 Mata kugeza 6 Gicurasi 2011 ku kibaya kigana mumajyepfo y'igishanga cya Migina). Ubu bwiganze bw'amazi mu migezi abanje kunyura mubutaka (cg munda y'isi) bujyanye neza neza n'ibipimo biri hasi byagaragajwe higwa isano riri hagati y'amazi atemba mu migezi n'imvura yaguye (runoff coefficient) nkuko biri hagati ya 16.7% na 44.5% kuri iyo myuzure yombi yakoreweho ubu bushakashatsi. Ku bw'ibyo rero, twanzuye ko amazi menshi aturuka mu butaka (mu nda y'isi), ndetse yiyongera cyane cyane mu bihe by'imvura akinjira mu migezi, ariyo atuma imigezi yo mu kibaya cy'igishanga cya Migina ahora atemba ndetse no mugihe cy'izuba ntakamemo.

Kuri icyo kibaya nanone cy'igishanga cya Migina, hakoreshejwe ubundi buhanga ariko bworoheje (simple rational method) bwita kubuso bw'ikibaya (area correction) mu guteganya amazi menshi ashoboka yatemba mu umugezi uri mu kibaya cy'igishanga cya Migina nawo witwa Migina. Ibyavuye muri ubu bushakashatsi nuko ikoreshwa ry'ubutaka mu buhinzi aririyo ryigaragaza cyane muri iki kibaya (kugipimo cya 92.5%). Ibyavuyemo kandi bigaragaza ko isano riri hagati y'amazi atemba mu mugezi n'imvura yaguye ari 25% muri rusange, naho igihe imvura yaguye kure hashoboka mu kibaya ishobora kumara itemba kugira ngo yinjire mu mugezi kingana n'amasaha 3 n'igice nyuma yo kuhagwa kw'imvura. Naho ubwinshi bushoboka bwo hejuru bwagaragaye mu itemba ry'amazi y'umugezi wa Migina (hafatiwe mu gihe cy'imyaka 10 cyisubiramo, 10 years return period) bugera kungano ya metero kibe 16 mu isogonda (16 m^3 s^{-1}).

Nyuma yubu bushakashatsi, twifuje kugenzura uko amazi atemba nyuma yo kugwa kw'imvura n'ingano y'umutungo kamere w'amazi ari kuri buri kibaya mu bibaya 5 biri mu kibaya cy'igishanga cya Migina. Kugira ngo tubigereho twakoresheje ubuhanga bwimbitse (bwa semi-distributed conceptual hydrological model) bwitwa HEC-HMS modelling (Hydrolologic Engeneering Centre- the Hydrologic Modelling System). Ubu bushakashatsi bwimbitse bwari butegerejweho kuzifashishwa nk'igikoresho cy'ingenzi mu buryo bw'igenamigambi mu bijyanye n'umutungo kamere w'amazi ndetse no gufata ibyemezo birambye byo kubungabunga no kubyaza umusaruro iki gishanga. Ubu buryo bwatoranyijwe kubera ubushobozi bwihariye bufite mu gusesengura imihindagurikire iranga intera y'amazi yinjira natinjira mu butaka, imyorohere y'imikoreshereze yabwo, ibyo busaba bike (imibare y'ibipimo) no kuba porogaramu yabwo iboneka itaguzwe.

Ubu buryo bwitwa HEC-HMS igice cyabwo cya 3.5 (HEC-HMS 3.5 model) bwakoreshejwe bushyizwe hamwe n'ibara (kubara) ry'ubuhehere bw'ubutaka, unit hydrograph, linear reservoir, kumenya ingano y'amazi yo kukigero cyo hasi (baseflow) n'ubundi buryo bw'imikorere bwa Muskingum-Cunge (river routing). Twakoresheje imibare y'ibipimo igaragaza imigwire y'imvura byo ku masitasiyo 12, n'imibare igaragaza itemba ry'amazi mu migezi byafatiwe ku masitasiyo 5. Ibi bipimo byose byubatse mu kibaya cyose cy'igishanga cya Migina harimo na amasitasiyo 2 agaragaza imihindakurikire y'izuba (pan evaporation) ndetse na amasitasiyo 11 (piezometers) apima aho amazi yo munda y'isi aherereye (GW Table) ndetse nuko ahindagurika (GW variation). Iyi mibare y'ibipimo ikaba yarakusanyijwe nk'igice cy'iyi nyigo mu gihe kingana n'imyaka ibiri (Gicurasi 2009 na Kamena 2011). Ikibaya cyose cyagabanyijwemo ibibaya bito 5 (Munyazi-Rwabuye, Mukura, Akagera, Cyihene-Kansi na Migina y'amajyepfo), buri cyose gihagarariwe na kimwe mu bipimo by'itemba ry'amazi (river gauging station) kumpera z'ikibaya.

Ibice bigize HEC-HMS model (model parameters) byahawe ingano, buri kimwe ukwacyo kuri buri gace cy'ikibaya hifashishijwe imibare y'ibipimo by'itemba ry'amazi byitegerejwe. Ibyavuye mu itangangano (model calibration) hasanzwe byakwemerwa ku masitasiyo 4 nyuma y'igenzura ryakoreshejwe ryitwa Nash-Sutcliffe model efficiency index (*NS*) ryerekanye ubwiza bw'imibare y'ibipimo bya buri munsi byafashwe bugera kuri 65% ku gipimo kiri ku mugezi aho ikibaya cyose kirangirira (Migina outlet station). Bitewe n'ibura ry'imibare y'ibipimo ihagije kandi yizewe yafashwe mu gihe kirebire, kugenzura ubwiza bw'ubu buryo ntabwo bwakoreshejwe (model validation (e.g. split sample test)). Gusa, twakoresheje ibyagaragajwe nuburyo bwa taraseri (tracer method) bushingiye mukugabanyamo ibice amazi yo mu mugezi (hydrograph separation). Ibi byakozwe hagamijwe kugereranya ibyagaragajwe n'ikoreshwa ry'ubuhanga bwimbitse (HEC-HMS model) mu bijyanye nuko amazi atemba yigabanyamo ibice kugirango agere mu migezi (runoff components).

Ibyavuye muri ubu bushakashatsi bwimbitse byagaragaje ko ubu buhanga bwakoze neza cyane mukugereranya ingano yose y'amazi yatembye mu mugezi (total flow volume). Bwanagaragaje igihe amazi yari menshi kuruta ibindi bihe mu mugezi kuburyo ashobora guteza imyuzure (peak flow) n'igihe yabereye menshi (timing), tutibagiwe igice cy'imvura cyahitaga kijya mu mugezi ntahandi inyuze atembeye gusa kubutaka (direct runoff). Ubushakashatsi kandi bwanagaragaje n'amazi yatembaga mu mugezi ntamvura yaguye aturutse munda y'isi (baseflow). Muri ubu bushakashatsi kandi twabonye ubusumbane (ukutangana) bukabije mu bice byaho amazi yibika iyo imvura imaze kugwa (urugero: ububiko bw'amazi yo mu butaka twita mu nda y'isi) n' ubusumbane mubice bigize itemba ry'amazi y'imvura (runoff components) muri buri kibaya uko ari bitanu. Ibi bikaba aribyo byaduhaye isura igaragaza itandukaniro ry'uruherererekane rw'inzira z'amasoko y'amazi kuri buri kibaya cy'igishanga n'ikindi bitewe n'imiterere yacyo. Kubwibyo, twanzuye tuvuga ko ubwo busumbane bugaragaza impamvu yo kwita ku miterere ya buri kibaya cy'igishanga ukwacyo, niba ibyo bipimo tubonye (parameters) ndetse n'ibice by'uruherererekane rw'amazi (components of water cycle) bigomba gushingirwaho mu gufata ibyemezo n'igenamigambi ku mutungo kamere w'amazi wo muri iki gishanga.

Ubumenyi buvuye muri ubu bushakashatsi buzashingirwaho n'abafata ibyemezo mu gushyiraho politiki n'ingamba zihamye z'igihugu zirebana n'igenamigambi n'imicungire iboneye y'umutungo kamere w'amazi. Ubumenyi bwagaragajwe muri ubu bushakashatsi buzakoreshwa no ku bindi bibaya by'inzuzi n'ibishanga by'u Rwanda kugira ngo bigire uruhare mu kugera ku cyerekezo cya 2020 u Rwanda rwihaye no mu ntego z'iterambere z'ikinyagihumbi (MDGs). Iri menyekana ry'umutungo kamere w'amazi dufite mu bibaya by'inzuzi no mu bishanga biri mu Rwanda, twizeye ko rizagabanya ikoreshwa rya cyane (ihihibikanya) ry'umutungo kamere w'amazi, kuko ibishanga bizajya bikoreshwa neza, cyane ko ubuhinzi (bukorerwa cyane bubishanga) bwihariye ibirenga 70% by'amazi yose akoreshwa ku rwego rw'igihugu.

Omar Munyaneza
UNESCO-IHE, i Delft, mu Buholandi
Mata 2014

....

The summary in Kinyarwanda was translated by Abuba Selemani and my brother Said Sibomana, Kigali, Rwanda

REFERENCES

Abbott, M.B., Bathurst, J.C., Cunge, J.A., O'Connell, P.E. and Rasmussen, J., 1986. An introduction to the European Hydrological System - Systeme Hydrologique Europeen, "SHE", 2: Structure of a physically-based, distributed modelling system. *Journal of Hydrology*, 87:61-77.

Abdulla, F.A., Amayreh, J.A., Hossain, A.H., 2002. Single event watershed model for simulating runoff hydrograph in desert regions. *Water Resources Management*, 16: 221-238.

Abushandi, E.H., 2011. Rainfall-Runoff Modelling in Arid Areas. *PhD Thesis*, Faculty for Geosciences, Geotechnique and Mining of the Technische Universität Bergakademie Freiberg, Germany.

AGRAR- UND HYDROTECHNIK, 1993. Schéma directeur d'aménagement et mise en valeur des marains du bassin de la Migina. *AGRAR-UND HYDROTECHNIK GMBH*, Beratende Ingenieure.Consulting Engineers, Essen, Germany.

Al-Adamat, R., Diabat, A., Shatnawi, G., 2010. Combining GIS with multicriteria decision making for silting water harvesting ponds in Northern Jordan. *Journal of Arid Environments*, 74: 1471-1477.

Arbind, K.V., Madan, K.J. and Rajesh, K.M., 2010. Evaluation of HEC-HMS and WEPP for simulating watershed runoff using remote sensing and geographical information system. *Paddy and Water Environment*, 8(2): 131-144. DOI: 10.1007/s10333-009-0192-8.

Anderson, J.E., Shiau, S.-Y., Harvey, K.D., 1992. Preliminary investigation of trend/patterns in surface water characteristics and climate variations. In: Kite, G.W., Harvey, K.D. (Eds.). *Using Hydrometric Data to Detect and Monitor Climatic Change*. Proceedings of NHRI Workshop No.8, *National Hydrology Research Institute, Saskatoon*, SK, pp. 189-201.

Ashagrie, A.G., De Laat, P.J.M., De Wit, M.J.M., Tu, M. and Uhlenbrook, S., 2006. Detecting the influence of land use changes on discharges and floods in the Meuse River Basin – the predictive power of a ninety-year rainfall-runoff relation? *Hydrol. Earth Syst. Sci.*, 10: 691-701.

Bedient, P.B. and W.C. Huber, 2002. Hydrology and Floodplain Analysis, 3rd Edition, Prentice Hall, NJ

Bergström, S. and Forsman, A., 1973. Development of a conceptual deterministic rainfall-runoff model. *Nordic Hydrology*, 4, pp. 147-170.

Beven, K. and Kirkby, M., 1979. A physically based, variable contributing area model of basin hydrology. *Hydrological Sciences Bulletin*, 24(1): 43-69.

Beven, K., 1989. Changing ideas in hydrology - the case of physically based models. *Journal of Hydrology*, 105, pp. 157-172.

Beven, K.J., 2000. Uniqueness of places and process representations in hydrological modelling. *Hydrology and Earth System Sciences*, 4: 203–213.

Beven, K., 2008. On doing better hydrological science. *Hydrol. Process.*, 22: 3549–3553.

Birkel, C., Tetzlaff, D., Dunn, S., and Soulsby, C., 2011. Using time domain and geographic source tracers to conceptualize streamflow generation processes in lumped rainfall-runoff models. *Water Resour. Res.*, 47: 14804-14831.

Blume, T., Zehe, E., Bronstert, A., 2007. Rainfall-runoff response, event-based runoff coefficients and hydrograph separation. *Hydrological Sciences Journal*, 52(5): 843-862.

Bonifacio, R., and Grimes, D.I.F., 1998. Drought and flood warning in southern Africa. *IDNDR Flagship Programme – Forecasts and Warnings*, UK National Coordination Committed for the IDNDR, Thomas Telford, London.Bouabid R., Elalaoui C.A., 2010. Impact of climate change on water resources in Moro cco: The case of Sebou Basin. *Séminaires Méditerranéens*, 9: 57- 62.

Boyogueno, S.H., Mbessa, M., and Tatietse, T.T., 2012. Prediction of Flow-Rate of Sanaga Basin in Cameroon Using HEC-HMS Hydrological System: Application to the Djerem Sub-Basin at Mbakaou. *Energy and Environment Research*, 2(1), ISSN 1927-0569 E-ISSN 1927-0577.

Burch, G.J., Bath, R.K., Moore, I.D. and O'Loughlin, E.M., 1987. Comparative hydrological behaviour of forested and cleared catchments in southeastern Australia. *J. Hydrol.*, 90:19–42.

Burn, D.H., Cunderlik, J.M., and Pietroniro, A., 2004. Hydrological trends and variability in the Liard River basin. *Hydrological Sciences Journal*, 49(1): 53-67.

Burn, D. H., Mansour, R., Zhang, K. and Whitfield, P. H., 2011. Trends and Variability in Extreme Rainfall Events in British Columbia, Canada. *Water Resour. J.*, 36: 67–82, doi:10.4296/cwrj3601067.

Burnash, R.J.C., Ferral, R.L. and McGuire, R.A., 1973. A generalized streamflow simulation system, conceptual modelling for digital computers. *Report by the Joint Federal State River Forecasting Centre*, Sacramento, CA, USA.

Burns, D. A., 2002. Stormflow-hydrograph separation based on isotopes: the thrill is gone-what's next? *Hydrol. Process.*, 16: 1515–1517.

Butler, D. and Davies, W.J., 2004. *Urban Drainage*, 2nd Ed., Spon Press, 11 New Fetter Lane, London.

Buttle, J. M., 1994. Isotope hydrograph separations and rapid delivery of pre-event water from drainage basins. *Progress in Physical Geography*, 18, 16-41.

Buytaert, W., Celleri, R., Willems, P., Bièvre, B.D., Wyseure, G., 2006. Spatial and temporal rainfall variability in mountainous areas: a case study from the south Ecuadorian Andes. *J. Hydrol.* 329 (3-4): 413–421.

Capell, R., Tetzlaff, D., and Soulsby, C., 2012. Can time domain and source area tracers reduce uncertainty in rainfall-runoff models in larger heterogeneous catchments? *Water Resour. Res.*, 48, W09544, doi:10.1029/2011WR011543, 14831.

CIA World Factbook, 2012, Rwanda: Economy. U.S. Dept. of State Country Background Notes.

Chen, H., Guo, S., Xu, C. and Singh, V.P., 2007. Historical temporal trends of hydro-climatic variables and runoff response to climate variability and their relevance in water resource management in the Hanjiang basin. *J. Hydrol.*, 344: 171-184.

Chingombe, W., Gutierrez, J.E., Pedzisai, E. and Siziba, E., 2005. A Study Of Hydrological Trends And Variability Of Upper Mazowe Catchment-Zimbabwe. *Journal of Sustainable Development in Africa*. Volume 7 #1. Spring, 17p.

Christopher A.J. and Yung A.C., 2001. The Use of HEC-GeoHMS and HEC-HMS to perform Grid-based Hydrologic Analysis of a Watershed, ASFPM Conference.

Chow, V.T., Maidment, D.R. and Mays, L.W., 1988. Applied hydrology, *McGraw-Hill*, Singapore, 99-126.

Chu, X. and Steinman, A.D., 2009. Event and Continuous Hydrologic Modeling with HEC-HMS, *Journal of Irrigation and Drainage Engineering*, ASCE 135: 119-124.

Clark, I. and Fritz, P., 1997. Environmental Isotopes in Hydrology, *CRC Press*, New York, USA, 1355–1356.

Clarke, R., 1973. A review of some mathematical models used in hydrology, with observations on their calibration and use. *Journal of hydrology*, 19(1): 1-20.

Collins, J.M., 2011. TemperatureVariability over Africa. *Journal of Climate*, 24(14): 3649-3666.

Cras, A., Marc, V. and Travi, Y., 2007. Hydrological behaviour of sub-Mediterranean alpine headwater streams in a badlands environment. *Journal of Hydrology*, 339(3-4): 130-144.

Crawford, N. and Linsley, R., 1966. Digital simulation in hydrology: Stanford Watershed Model IV. *Technical Report No. 39, Department of Civil Engineering, Stanford University, Stanford, CA, USA.*

Cunderlik, J.M. and Simonovic, S. P., 2004. Calibration, verification and sensitivity analysis of the HEC-HMS hydrologic model. CFCAS project: Assessment of Water Resources Risk and Vulnerability to Changing Climatic Conditions. *Project report IV,* August 2004.

Cunderlik, J.M. and Simonovic, S.P., 2005. Hydrological extremes in a south-western Ontario river basin under future climate conditions, *Hydrological Sciences Journal,* 50: 631–654.

Czop, P., Kost, G., Sławik, D., and Wszołek, G., 2011. Formulation and identification of First-Principle Data-Driven models, *Journal of Achievements in Materials and Manufacturing Engineering,* 44(2): 179-186.

Davis, B., Winters, P., Carletto, G., Covarrubias, K., Qui~nones, E., Zezza, A., Stamoulis, K., Azzarri, C. and DiGiuseppe S., 2010. A cross-country comparison of rural income generating activities. *World Development,* 38(1): 48-63.

Dawod, M.G. and Koshak, A.N., 2011. Developing GIS-Based Unit Hydrographs for Flood Management in Makkah Metropolitan Area, Saudi Arabia. *Journal of Geographic Information System,* 3: 153-159.

De Brouwer, J.A.M., 1997. Determination of peak runoff. Chapter 10 In: ITC, 1997: The Integrated Land and Water Information System, Application Guide, ILWIS 2.1 for Windows. *International Institute for Aerospace Survey & Earth Sciences,* Enschede, The Netherlands.

De Laat, P.J.M., Savenije, H.H.G., 2002. Hydrology. Lecture notes. *UNESCO-IHE Institute for Water Education,* Delft, The Netherlands.

De Laat, P.J.M., 2006. Workshop on Hydrology. Lecture notes. *UNESCO-IHE Institute for Water Education,* Delft, The Netherlands.

Didszun, J. and Uhlenbrook, S., 2008. Scaling of dominant runoff generation processes: Nested catchments approach using multiple tracers. *Water Resour. Res.,* 44, W02410, doi:10.1029/2006WR005242.

Dong, X.H., Yao, Z.J., Chen, C.Y., 2007. Runoff variation and its response to precipitation in the source region of the Yellow River. *Resources Science,* 29(3): 67–73.

Du, J., Xie, S., Xu, Y., Chong-yu, X. and Vijay P.S., 2007. Development and testing of a simple physically-based distributed rainfall-runoff model for storm runoff simulation in humid forested basins. *Journal of Hydrology,* 336: 334–346.

Dushimire, H.A., 2007. Water balance of Lake Muhazi. *MSc thesis, National University of Rwanda,* Department of Civil Engineering, Butare, Rwanda, 78 pp.

EarthTrends, 2003. Water Resources and Freshwater Ecosystems in Rwanda. *EarthTrends Country Profiles,* Visited in November 2008, http://earthtrends.wri.org

Emerson, C.H., Welty, C. and Traver, R.G., 2003. Application of HEC-HMS to model the additive effects of multiple detention basins over a range of measured storm volumes. *Civil Engineering Database,* Part of world water & Environmental Resources Congress 2003 and Related Symposia.

Fadul, H.M. and Ali, I.A., 2009. Climate Variability of the Nile basin; with Reference to Sudan. *Regional NBCBN Conference on Flood management,* Nairobi, Kenya, 20-23 April 2009.

FAO, 1997. Irrigation potential in Africa: A basin approach. *FAO Land and Water Bulleting,* No.4, FAO, Rome.

FAO, 2005. Système d'information de la FAO sur l'eau et l'agriculture (in French), Information system for water and agriculture. *Food and Agriculture Organization (FAO) of the United Nation,* Rome, Italy.

FAO, 2006. Design model for catchment: Cultivated area ratio. *Food and Agriculture Organization (FAO) of the United Nation*, Natural Resources Management and Environment Department, Rome, Italy.

Fenicia, F., Savenije, H.H.G., Matgen, P., and Pfister, L., 2008a. Understanding catchment behavior through stepwise model concept improvement, *Water Resour. Res.*, 44, W01402, doi:10.1029/2006WR005563, 14812-14831.

Fenicia, F., McDonnell, J.J., and Savenije, H.H.G., 2008b. Learning from model improvement: On the contribution of complementary data to process understanding, *Water Resour. Res.*, 44, W06419, doi:10.1029/2007WR006386.

Fenicia, F., Savenije, H.H.G., and Winsemius, H.C., 2008c. Moving from model calibration towards process understanding, *Physics and Chemistry of the Earth*, 33: 1057–1060.

Fenicia, F., Kavetski, D., and Savenije, H.H.G., 2011. Elements of a flexible approach for conceptual modeling: Part 1. Motivation and theoretical development. *Water Resources Research*, 47, W11510, doi:10.1029/2010WR010174, 14809.FGDC., F.G.D.C., 1998. National Standard for Spatial Data Accuracy, In: *F.G.D.C. FGDC.* (Editor), Geospatial Positioning Accuracy Standards, pp. 3-11.

Fleming, G., 1975. Computer simulation techniques in hydrology. *Elsevier, New York*, NY, USA.

Fleming, M. and Neary, V., 2004. Continuous Hydrologic Modeling Study with the Hydrologic Modeling System, *Journal of Hydrologic Engineering*, ASCE 9: 175–183.

Gao, H., Hrachowitz, M., Fenicia, F., Gharari, S., and Savenije, H.H.G., 2013. Testing the realism of a topography driven model (FLEX-Topo) in the nested catchments of the Upper Heihe, China. *Hydrol. Earth Syst. Sci. Discuss.*, 10: 12663-12716.

GEF, 2006. Transboundary Agro-Ecosystem Management Programme for the Kagera River basin. *UNEP/EFO*, Sept 2006.

Gharari, S., Hrachowitz, M., Fenicia, F., Gao, H., and Savenije, H. H. G., 2013.Using expert knowledge to increase realism in environmental system models can dramatically reduce the need for calibration, *Hydrol. Earth Syst. Sci. Discuss.*, 10: 14801-14855.

Githui, F., Gitau, W., Mutua, F., and Bauwensa, W., 2009. Climate change impact on SWAT simulated stream flow in western Kenya. *International Journal of Climatology*, 29:1823–1834.

Graham, D.N. and M. B. Butts, 2005. Flexible, integrated watershed modelling with MIKE SHE. In Watershed Models, Eds. V.P. Singh & D.K. Frevert Pages 245-272, CRC Press. ISBN: 0849336090.

Grayson, R.B. and Blöschl, G., 2000. Spatial Patterns in Catchment Hydrology: Observations and Modelling. *Cambridge University*, Press. 404p.

Guinot, V. and Gourbesville, P., 2003. Calibration of physically based models: back to basics? *Journal of Hydroinformatics*, 5(4): 233-244.

Haan, C.T, Johnosn, H.P., Brakensiek, D.L. (Eds.), 1982. Hydrologic Modelling of Small Watersheds. *St. Joseph, Michigan: American Society of Agricultural Engineers.*

Hawkins, R.H., 1975. The importance of accurate curve numbers in estimation of storm runoff. *Water Resource Bulletin*, 11: 887-891.

HEC, 2011. Hydrologic Modelling System, HEC-HMS, User's Manual, Version 2.1. *Hydrologic Engineering Center (HEC), U.S. Army Corps of Engineering*, Davis, CA.

Helsel, D.R., Hirsch, R.M., 1992. Statistical methods in water resources. *Elsevier, Amsterdam*, 522 pp.

Higgins, J., 2005. The Radical Statistician. *Prentice Hall Publishing.*

Hirsch, R.W., Slack, J.R., Smith, R.A., 1982. Techniques of trend analysis for monthly water quality data. *Water Resources Management*, 22: 1159–1171.

Hitimana, J., Namara, A., Sengalama, T. and Nyirimana, J., 2006. Community-Based Natural Resource Management (CBNRM) Plan. Kinigi Area, Rwanda. *Report prepared for the International Gorilla Conservation Programme*, Kigali, Rwanda.

Hrachowitz, M., Bohte, R., Mul, M. L., Bogaard, T. A., Savenije, H. H. G., and Uhlenbrook, S., 2011. On the value of combined event runoff and tracer analysis to improve understanding of catchment functioning in a data-scarce semi-arid area. *Hydrol. Earth Syst. Sci.*, 15, 2007–2024.

Hrachowitz, M., Savenije, H.H.G., Bogaard, T.A., Tetzlaff, D., and Soulsby, C., 2013a. What can flux tracking teach us about water age distribution patterns and their temporal dynamics? *Hydrol. Earth Syst. Sci.*, 17: 533–564.

Hrachowitz, M., Savenije, H.H.G., Blöschl, G., McDonnell, J.J., Sivapalan, M., Pomeroy, J.W., Arheimer, B., Blume, T., Clark, M.P., Ehret, U., Fenicia, F., Freer, J.E., Gelfan, A., Gupta, H.V., Hughes, D.A., Hut, R.W., Montanari, A., Pande, S., Tetzlaff, D., Troch, P.A., Uhlenbrook, S., Wagener, T., Winsemius, H.C., Woods, R.A., Zehe, E., and Cudennec, C., 2013b. A decade of Predictions in Ungauged Basins (PUB)—a review. *Hydrological Sciences Journal*, 58(6): 1–58.

Hoeg, S., Uhlenbrook, S., and Leibundgut, C., 2000. Hydrograph separation in a mountainous catchment – combining hydrochemical and isotopic tracers. *Hydrol. Process.*, 14(7): 1199-1216.

Hu, Y., Maskey, S., Uhlenbrook, S., 2011a. Trends in temperature and precipitation extremes in the Yellow River source region, China. *Climatic Change*, 1-27, DOI: 10.1007/s10584-011-0056-2.

Hu, Y., Maskey, S., Uhlenbrook, S., and Zhao, H., 2011b. Streamflow trends and climate linkages in the source region of the Yellow River, China. *Hydrological Processes*, 25: 3399–3411.

IHA, 2001. User's manual, Indicators of Hydrologic Alteration. *The Nature Concervancy*.

IPCC (Intergovernmental Panel on Climate Change) (2001). Climate Change 2001: Impacts, Adaptation, and Vulnerability. Contribution of the Working Group II to the Third Assessment Report of the IPCC. *Cambridge University Press*, Cambridge, UK.

IPCC, 2007. Climate Change 2007. The Physical Science Basis. Contribution of WorkingGroup I to the Fourth Assessment Report of the Intergovernmental Panel on Climate Change, S. Solomon, D. Qin, M. Manning, Z. Chen, M. Marquis, K.B. Averyt, M. Tignor and H.L. Miller, Eds., *Cambridge University Press*, Cambridge, 996 pp.

Iroumé, A., Huber, A. and Schulz, K., 2005. Summer flows in experimental catchments with different forest covers, Chile. *J. Hydrol.*, 300(1/4): 300–313.

Jakeman, A.J. and Hornberger, G.M., 1993. How much complexity is warranted in a rainfall-runoff model, *Water Resources Research*, 29(8): 2637-2649.

James, A.L. and Roulet, N.T.,2009. Antecedent moisture conditions and catchment morphology as controls on spatial patterns of runoff generation in small forest catchments. *Journal of Hydrology*, 377: 351–366.Kabalisa V. P., 2006. Analyse contextuelle en matière de Gestion Intégrée des Ressources en Eau au Rwanda, *Document de Travail pour l'ONG Protos* (Rapport final).

Kabeya, N., Shimizu, A., Chann, S., Tsuboyama, Y., Nobuhiro, T., Keth, N., and Tamai, K., 2007. Stable isotope studies of rainfall and stream water in forest watershed in Kampong Thom, Cambodia. *Forest environments in the Mekong River Basin*, 1:125–134, doi:10.1007/978-4-431-46503-4_11.

Kaffas, K. and Hrissanthou, V., 2014. Application of a continuous rainfall-runoff model to the basin of Kosynthos river using the hydrologic software HEC-HMS. *Global NEST Journal*, 16(1): 188-203.

Kasanziki, Ch., 2008. Assessment of water quality of springs; study of some springs around Butare town: Gahenerezo, Rwasave, Agasharu and Mpazi, MSc Thesis in Water Resources and Environmental Management, *National University of Rwanda*, Butare, Rwanda.

Kashid, S.S., Ghosh, S., and Maity, R., 2010. Streamflow Prediction using Multi-Site Rainfall Obtained from Hydroclimatic Teleconnection, *Journal of Hydrology*, Elsevier, 395(1-2), 23-38.

Klemes, V., 1986. Operational testing of hydrological simulation models. *Hydrol. Sci. J.*, 31: 13-24.

Kendall, M.G., 1975. Rank Correlation Methods. *Oxford Univ. Press*, New York, 202 pp.

Kennedy, E.J., 1984. Discharge rating at gauging stations, *U.S. Geological Survey Techniques of Water Resources Investigations*, Book 3, Chapter A10, 59pp.

Kennedy, V. C., Kendall, C., Zelleweger, G. W., Wyerman, T. A., and Avanzino, R. J., 1986. Determination of the components of stormflow using water chemistry and environmental isotopes, Mattole River basin, California. *Jour. of Hydrol.*, 84: 107-140.

Kente, L., 2011. Overview of water resources in Rwanda. *International Hydrological Programme (IHP) Conference, Rwanda National Commission for UNESCO (RNCU)*, Laico Umubano Hotel, Kigali, Rwanda, 28-29 April 2011. p7-10.

Kipkemboi, J., 2005. Utilization of wetlands through integration of finger ponds into riparian farming systems in East Africa. PhD thesis, *UNESCO-IHE Institute for Water Education*, Delft, The Netherlands.

Kirpich, Z.P., 1940. Time of concentration of small agricultural watersheds. *Civil Engineering*, 10(6), 362. The original source for the Kirpich equation.

Kuichling, E., 1889. The relation between the rainfall and the discharge of sewers in populous districts. *Transactions, American Society of Civil Engineers*, 20: 1-56.

Kyaruzi, A., Hailu, Z., Mngodo, R.J., Mkandi, S.H., Ntungumburanye, G.g Kizza, M. and Malisa, J., 2005. Regional power integration in hydropower. Group II Scientific Report, *Nile Basin Capacity Building Network*, Cairo, Egypt, p114.

Ladouche, B., Probst, A., Viville, D., Idir, S., Baque, D., Loubet, M., Probst, L.J., and Bariac, T., 2001. Hydrograph separation using isotopic, chemical and hydrological approaches (Strengbach catchment, France). *Journal of Hydrology*, 242: 255-274.

Langousis, A., 2005. The Area Reduction Factor (ARF): a multifractal analysis. Massachusetts *Institute of technology, Cambridge*.

Larsen, I.J., MacDonald, L.H., Brown, E., Rough, D., Welsh, M.J., Pietraszek, J.H., Libohova, Z., and Schaffrath, K., 2007. Causes of post-fire runoff and erosion: the roles of soil water repellency, surface cover, and soil sealing, Department of Forest, Rangeland, and Watershed Stewardship, *Colorado State University*, Fort Collins, Colorado.

Legates, D.R., 1999. Evaluating the Use of 'Goodness of Fit' Measures in Hydrologic and Hydroclimatic Model Validation. *Water Resources Research*, 35:233-241.

Leibundgut, Ch., 1998. Tracer-based assessment of vulnerability in mountainous headwaters. IAHS Publication, No 248, p. 317.Ley, R., Casper, M.C., Hellebrand, H. and Merz, R., 2011. Catchment classification by runoff behavior with self-organizing maps (SOM). Hydrol. Earth Syst. Sci., 15: 2947–2962.

Loukas, A. and Vasiliades, L., 2014. Streamflow simulation methods for ungauged and poorly gauged watersheds. *Nat. Hazards Earth Syst. Sci. Discuss.*, 2: 1033–1092.Love, D., Uhlenbrook, S., Twomlow, S., and Van der Zaag, P., 2010. Changing hydro-climatic and discharge patterns in the northern Limpopo Basin, Zimbabwe. ISSN 0378-4738, Water SA (Online) Vol. 36 No. 3, Pretoria, Apr., 2010.Mann, H.B., 1945. *Non-parametric test against trend*. Econometrika, 13: 245– 259.

Ma, Q., 2012. Analyzing impacts of land use change on hydrological regimes of Haihe river basin in China. MSc thesis, *UNESCO-IHE Institute of Water Education*, Delft, The Netherlands.

Maathius, B.H.P., Gieske, A.S.M., Retsios, V., Leeuwen, B., Maanaerts, C.M. and Hendrikse, J.H.M., 2006. Meteosat-8: From temperature to rainfall. *ISPRS Technical Commission VII*, Mid Term Symposium 8-11 May 2006.

Magoma, D.M., 2009. Hydrological Modeling Using SWAT in Wetland Catchments: The Case Study of Rugezi Watershed in Rwanda, First annual Nile Basin research conference, Dar es Salaam, Tanzania.Mann, H.B., 1945. Non-parametric test against trend. Econometrika, 13: 245– 259.

Marchi, L., Borga, M., Preciso, E., and Gaume, E., 2010. Characterisation of selected extreme flash floods in Europe and implications for flood risk management; *Journal of Hydrology*, 394(1-2): 118–133.

Martin, P.H., LeBoeuf, E.J., Dobbins, J.P., Daniel, E.B. and Abkowitz, M.D., 2005. Interfacing GIS with water resource models: A state-of-the-art review. *Journal of the American Water Resources Association*, 41:1471-1487.

Masih, I., Uhlenbrook, S., Maskey, S. and Smakhtin, V., 2010. Streamflow trends and climate linkages in the Zagros Mountains, Iran. *Climatic Change*, 104: 317–338.

Masih, I., Maskey, S., Uhlenbrook, S. and Smakhtin, V., 2011. Assessing the impact of areal precipitation input on streamflow simulations using the SWAT model. *Journal of the American Water Resources Association* (JAWRA) 47(1): 179-195.

Masih, I., 2011. Understanding hydrological variability for improved water management in the semi-arid Karkheh basin, Iran. PhD thesis, *UNESCO-IHE Institute for Water Education*, Delft, The Netherlands, ISBN: 978-0-415-68981-6 (Taylor & Francis Group).

Mazvimavi, D., 2003. Estimation of Flow Characteristics of Ungauged Catchments: Case Study in Zimbabwe. PhD thesis, *Wageningen University*, The Netherlands, 155 pp.

McCuen, R.H., Knight, Z., and Cutter, A.G., 2006. Evaluation of the Nash-Sutcliffe Efficiency Index. *J. Hydrol. Eng.*, 11(6): 597–602.McDonnell, J.J, Bonell, M., Stewart, K.M., and Pearce, J.A., 1990. Deuterium Variations in Storm Rainfall: Implications for Stream Hydrograph Separation. *Water Resour. Res.*, 26(3): 455-458.

McDonnell, J.J., 2003. Where does water go when it rains? Moving beyond the variable source area concept of rainfall–runoff response. *Hydrological Processes*, 17: 1869–1875.

Merwade, V., 2007. Hydrologic Modelling using HEC-HMS. School of Civil Engineering. *Purdue University.*

Meteo Rwanda, 2002. Rainfall Atlas, Rwanda Meteorological Office, Ministry of Infrastructure, Kigali, Rwanda, March 2002.

Miao, C.Y., Duan, Q.Y., Sun, Q.H., Li, J.D., 2013. Evaluation and application of Bayesian Multi-model estimation in temperature simulations. *Progress in physical geography.* doi:10.1177/0309133313494961.

Mihalik, E.N., 2007. Rainfall Runoff Processes, College of Engineering, *Purdue University.*

Mikova, K., Wali , U.G. and Nhapi, I., 2009. Predicting the influence of climate change on rainfall dynamic in Rwanda, Regional NBCBN Conference, Kigali, Rwanda, 01-03 December 2009.

Mikova, K., Wali, U.G., and Nhapi, I., 2010. Infilling of Missing Rainfall Data from a Long Term Monitoring Records. International Journal of Ecology & Development, 16: 89-99.

MINIRENA, 2013. Rwanda: Government Moves to Upgrade Meteorology Facilities, Kigali, Rwanda.

MINIPLAN, 2002. Rencement Général de la Population et de l'Habitat. *Ministry of Planning, Republic of Rwanda*, Kigali Rwanda.

MINITERE, 2005., Technical Assistance for the Preparation of the National Water Resources Management Project, *Ministère des Terres, de l'Environnement, des Forêts, de l'Eau et des Mines (MINITERE)*, Kigali, Rwanda.

MINITERE, 2006. National Adaptation Programmes of Action (NAPA) To Climate Change. *Ministry Of Lands, Environment, Forestry, Water and Mines (MINITERE)*, Kigali, Rwanda.

MINITERE, 2007. Rapport d'inventaire des ressources ligneuses au Rwanda. Vol. 2, *Ministère des Terres, de l'Environnement, des Forêts, de l'Eau et des Mines (MINITERE) and the National Agricultural Research Institute (ISAR)*, Kigali, Rwanda.

Moeyersons, J., 1991. Ravine formation on steep slopes: forward versus regressive erosion. Some case studies from Rwanda, Catena, 18, 309-324.

Molden, D., Oweis, T.Y. and Pasquale, S., 2007. Pathways for increasing agricultural productivity. In Molden, D. (ed) Water for food, water for life: a comprehensive assessment of water management in agriculture. *Earthscan, London*, pp 279-310.

Mul, M.L., Mutiibwa, K.R., Uhlenbrook, S., and Savenije, H.G.H., 2008a. Hydrograph separation using hydrochemical tracers in the Makanya catchment, Tanzania. *Physics and Chemistry of the Earth*, (33), 151–156.

Mul, M.L., Savenije, H.G. and Uhlenbrook, S., 2008b. Spatial rainfall variability and runoff response during an extreme event in a semi-arid catchment in the South Pare Mountains, Tanzania. *Phys. Chem. Earth*, Vol. 3, 1-2, 151-156.

Mul, M. L., 2009. Understanding hydrological processes in an ungauged catchment in Sub-Saharan Africa. PhD thesis, *UNESCO-IHE Institute for Water Education*, Delft, The Netherlands, ISBN: 978-0-415-54956-1 (Taylor & Francis Group).

Munyaneza, O., Wenninger, J., and Uhlenbrook, S., 2012a. Identification of runoff generation processes using hydrometric and tracer methods in a meso-scale catchment in Rwanda, *Hydrol. Earth Syst. Sci.,* 16: 1991–2004.

Munyaneza, O., Nizeyimana, G., Nsengimana, H., Uzayisenga, Chr., Uwimpuhwe, Ch. and Nduwayezu, J.B., 2012b. *Surface Water Resources Assessment in the Rwasave Marshland, southern Rwanda*. Nile Water Sci. Eng. J., 5(2): 58-70

Munyaneza, O., Ndayisaba, C., Wali, U.G., Mulungu, M.M.D., and Dulo, O.S., 2011a. Integrated Flood and Drought Management for Sustainable Development in the Kagera River Basin. *Nile Water Sci. Eng. J.,* 4(1): 60-70.

Munyaneza, O., Ufiteyezu, F., Wali, U.G. and Uhlenbrook, S., 2011b. A simple Method to Predict River Flows in the Agricultural Migina Catchment in Rwanda. *Nile Water Sci. Eng. J.,* 4(2): 24-36.

Munyaneza, O., Uhlenbrook, S., Wenninger, J., van den Berg, H., Bolt, H. R., Wali, G. U., and Maskey, S., 2010. Setup of a Hydrological Instrumentation Network in a Meso-Scale Catchment- the case of the Migina Catchment, Southern Rwanda. *Nile Water Sci. Eng. J.,* 3(1): 61-70.

Munyaneza, O., Wali, U.G, Uhlenbrook, S., Maskey, S. and Mlotha, M.J, 2009a. Water Level Monitoring using Radar Remote Sensing Data: Application to Lake Kivu, Central Africa, *Journal of Physics and Chemistry of the Earth*, 34: 722-728.

Munyaneza, O., Uhlenbrook, S., Maskey, S., Wali, U. G. and Wenninger, J., 2009b. Hydrological and climatic data availability and preliminary analysis in Rwanda, Proceedings of Hydrology session, *10th International WaterNet/WARFSA/GWP-SA Symposium*, Entebbe, Uganda, 28-30 October 2009.

Musoni, J.P., Wali, U.G. and Munyaneza, O., 2009. Runoff coefficient classification on Nyabugogo catchment, Proceedings of the Hydrology, *10th International WaterNet/WARFSA/GWP-SA Symposium*, Entebbe, Uganda, 28-30 October 2009.

Mutabazi, A., Kabalisa, V.P., Kayitare, R.E., Hakizimana, E., Dusabeyezu, S. and Rwema, E., 2004. Generation and application of climate information, products and services for disaster preparedness and sustainable development in Rwanda. *Rwanda Meteorological Service*, Kigali, Rwanda, June 2004.

Nahayo, D., 2008. Feasible Solutions for an Improved Watershed Management in Sloping Areas, Rwanda, Proceedings of Water and Land session, *9th WaterNet/WARFSA/GWP-SA Symposium*, Johannesburg, South Africa, 29–31 October, 2008.

Nahayo, D., Wali, U.G. and Anyemedu, F.O.K., 2010. Irrigation practices and water conservation opportunities in Migina marshlands. *International Journal of Ecology & Development*, 16, 100-112.

Nash, J.E. and Sutcliffe, J.V., 1970. River flow forecasting through conceptual models part I – A discussion of principles, *Journal of Hydrology*, 10(3): 282-290.

Neary, V.S., Habib, E. and Fleming, M., 2004. Hydrologic Modeling with NEXRAD Precipitation in Middle Tennessee, *Journal of Hydrologic Engineering*, ASCE, 9: 339–349.

NELSAP, 2007 Natural Resources Management and Development, NELSAP Rwanda, Visited in November 2008, available at: http://web.worldbank.org/Rwanda, last access: 25 November 2008.Nielsen, S. and Hansen, E., 1973. Numerical simulation of the rainfall-runoff process on a daily basis. *Nordic Hydrology*, 4(3): 171–190.

NISR, 2012. The Fourth Rwanda Population and Housing Census (RPHC4), *National Institute of Statistic of Rwanda* (NISR), Kigali, Rwanda.

NWRMP, 2008. Rehabilitation of hydrological stations in Rwanda, *National Water Resources Management Project, Ministry of Natural Resources*, Kigali, Rwanda.

O'Loughlin, G., Huber, W. and Chocat, B., 1999. Rainfall-runoff processes and modelling, *Journal of Hydraulic Research*, 34(6): 733-751.

Oyebande, L., 2001. Water problems in Africa-how can sciences help? Hydrological *Sciences Journal*, 46(6): 947-961.

Pajunen, H., 1996. Mires as late Quaternary accumulation basins in Rwanda and Burundi, Central Africa. *Geological Survey of Finland*, bulletin 384, Espoo

PGNRE, 2005a. Assistance technique a la preparation du projet de gestion nationale des ressources en eau. Composante B: Connaissance et gestion des données sur l'eau. Rapport Final Définitif. *Ministry of Natural Resources (MINIRENA)*, Kigali, Rwanda.

PGNRE, 2005b. Projet de gestion nationale des ressources en eau (PGNRE). Composante D: Etudes technique. Rapport General. *Ministry of Natural Resources (MINIRENA)*, Kigali, Rwanda.

Peters, N.E., 1994. Biogeochemistry of Small Catchments: A Tool for Environmental Research. Edited by B. Moldan and J. Cern. *Published by John Wiley & Sons Ltd.*

Pettitt, A.N., 1979. A non-parametric approach to the change-point problem. *Appl. Stat.*, 28: 126-135.

Pilgrim, D. H., Chapman, T. G., Doran, D. G., 1988. Problems of Rainfall-Runoff Modelling in Arid and Semiarid Regions. Hydrological Sciences Journal, 33: 379-400.

Pinder, G. F. and Jones, J. F., 1969. Determination of the groundwater component of peak discharge from the chemistry of total runoff. *Water Resour. Res.*, 5(2), 438-445.

Price, R.K., 2006. The growth and significance of hydroinformatics. In: D.W. Knight and A.Y. Shamseldin (eds.), *River Basin Modelling for Flood Risk Mitigation*, Taylor & Francis, London, UK.

PRSP, 2007. Environmental Sustainability in Rwanda's Economic Development and Poverty Eradication Strategies: Towards Mainstreaming Environment in The EDPRS. *Rwanda's Poverty Reduction Strategy Paper*, Kigali-Rwanda.Radmanesh, F., Por Hemat, J., Behnia, A., Khond, A. and Ali Mohamad, B., 2006. Calibration and assessment of HEC- HMS model in Roodzard watershed, *17th International conference of river engineering*, University of Shahid Chamran, Ahvaz.

Refsgaard, J.C., 1996. Terminology, modelling protocol and classification of hydrological model codes. In: M.B. Abbott and J.C. Refsgaard (eds.), Distributed Hydrological Modelling, Kluwer Academic Publishers, Dordrecht, The Netherlands, pp. 17-39.

Reinelt, L.E., Velikanje, J. and Bell, E.J., 1991. Development and Application of a Geographic Information-System for Wetland Watershed Analysis. *Computers Environment and Urban Systems*, 15: 239-251.

REMA, 2009. Economic impact of climate change in Rwanda. Kigali, Rwanda.

REMA, 2009b. Rwanda State of Environment and Outlook Report, *Rwanda Environment Management Authority (REMA)*, Kigali, Rwanda.

Renno, C. D., Nobre, A. D., Cuartas, L. A., Soares, J. V., Hodnett, M. G., Tomasella, J., and Waterloo, M. J., 2008. HAND, a new terrain descriptor using SRTM-DEM: Mapping terra-firme rainforest environments in Amazonia. *Remote Sens. Environ.*, 112: 3469–3481.

Richey, G.D., McDonnell, J.J., Erbe, W.M., and Hurd, M.T., 1998. Hydrograph separation based on chemical and isotopic concentrations: a critical appraisal of published studies from New Zealand, North America and Europe. *Journal of Hydrology* (NZ) 37(2), 95-111.

Richter, B.D, Baumgartner, J.V., Wigington, R. and Braun. D.P., 1997. How much water does a river need? *Freshwater Biology*, 37: 231-249.

Richter, B.D., Baumgartner, J.V., Powell, J. and Braun, D.P., 2009. A Method for Assessing Hydrologic Alteration within Ecosystems. The Nature Conservancy, Boulder, Colorado.

RIWSP, 2012a. Review of the National Hydrological Service in Rwanda (R. Venneker, J. Wenninger), RIWSP-No. 2.2.1/02, *Rwanda Integrated Water Security Program*, Kigali, Rwanda.

RIWSP, 2012b. *Assessment of available water resources data and information from selected hydrometric stations in Rwanda* (R. Venneker, J. Wenninger), Internal report to the Rwanda Natural Resources Authority, RIWSP-No. 2.2.1/03 Rwanda Integrated Water Security Program, Kigali, Rwanda.

RIWSP, 2012c. *RNRA hydrometric archive - Data conversion and recovery report*, Version 3 (R. Venneker, J. Wenninger), Internal report to the Rwanda Natural Resources Authority, RIWSP-No. 2.2.1/04 Rwanda Integrated Water Security Program, Kigali, Rwanda.

RIWSP, 2012d. *RNRA hydrometric observing network - Observing station details*, Version 2 (R. Venneker, J. Wenninger), Internal report to the Rwanda Natural Resources Authority, RIWSP-No. 2.2.1/05 Rwanda Integrated Water Security Program, Kigali, Rwanda.

RNRA, 2012. Review of the National Hydrological Services, *Rwanda Natural Resources Authority*, Kigali, Rwanda.

Roy, D., Begam, S., Ghosh, S. and Jana, S., 2013. Calibration and validation of HEC-HMS model for a river basin in Eastern India. ARPN Journal of Engineering and Applied Sciences, 8(1): 847-. ISSN 1819-6608.

Ruelland, D., Ardoin-Bardin, S., Billen, G. and Servat, E., 2008. Sensitivity of a lumped and semi-distributed hydrological model to several methods of rainfall interpolation on a large basin in West Africa. *J. Hydrol.*, 361(1-2), 96-117.

Safari, B., 2012. Trend Analysis of the Mean Annual Temperature in Rwanda during the Last Fifty Two Years. *Journal of Environmental Protection*, 3: 538-551.

Sang, J.K., 2005. Modeling the impact of changes in land use, climate and reservoir storage on flooding in the Nyando basin. M.Sc. Thesis, *Jomo Kenyatta University of Agriculture and Technology*, Kenya.

Sardoii, E. R., Rostami, N., Sigaroudi, Sh. K. and Taheri, S., 2012. Calibration of loss estimation methods in HEC-HMS for simulation of surface runoff (Case Study: Amirkabir Dam Watershed, Iran). *Adv. Environ. Biol.*, 6(1): 343-348.

Savenije, H.H.G., 2009. The art of hydrology. *Hydrology and Earth System Sciences*. 13: 157–161.

Savenije, H.H.G., 2010. Topography driven conceptual modeling, FLEX-Topo. *Hydrology and Earth System Sciences*, 14: 2681–2692

Savenije, H.H.G. and Sivapalan, M., 2013. PUB in practice: case studies. Chapter 11 In: G. Blöschl, *et al.*, eds. Runoff prediction in ungauged basins: synthesis across processes, places and scales. Cambridge: *Cambridge University Press*, 270–360.

Schaefli, B. and Gupta, H. V., 2007. Do Nash values have value?. *Hydrol. Process.*, 21: 2075–2080. doi: 10.1002/hyp.6825.

Scharffenberg, W. and Fleming, M., 2010. Hydrologic modelling system HEC-HMS v3.2 user's manual. USACE-HEC,Davis, USA.

Schwab, G.O., Fangmeier D.D., Elliot W.J. and Freveret R.K., 1993. Soil and water conservation engineering, *J. Wiley and sons.* New York. 507 pp.

Shadeed, S., 2008. Up To Date Hydrological Modelling in Arid and Semi-arid Catchment, the Case of Faria Catchment, West Bank, Palestine. Ph. D. Dissertation, *Freiburg University*, Germany.

Seibert, J., Bishop, K., Rodhe, A., and McDonnell, J. J., 2003. Groundwater dynamics along a hillslope: a test of the steady state hypothesis, Water Resour. Res., 39, 1014, doi:10.1029/2002WR001404, 14804-14821.

Sharif, M. and Burn, D., 2009. Detection of Linkages Between Extreme Flow Measures and Climate Indices. *World Academy of Science, Engineering and Technology*, 60: 871-876.

Shaw, M.E., 2004. Hydrology in Practice, Third edition, *Department of Civil Engineering, Imperial College of Science, Technology and Medicine*, ISBN 0748744487.

SHER, 2003. Etudes de faisabilite, Marais de la Migina. Rapport provisoire phase 2. *Groupement HYDROPLAN* Ingenieurs-Conseils, *Ministry of Agriculture*, Kigali, Rwanda.

SHER, 2004. Stremflows and climate database in Rwanda (PAGNRE). *Groupement HYDROPLAN Ingénieurs-Conseils s.a.*, Kigali, Rwanda.

Shetkar, R.V. and Mahesha, A., 2011. Tropical, Seasonal River Basin Development: Hydrogeological Analysis. *J. Hydrol. Eng.*, 16: 280-292; doi:10.1061/(ASCE)HE.1943-5584.0000328.

Shrestha, L.D., 2009. Uncertainty analysis in rainfall-runoff modelling: application of machine learning techniques. PhD Thesis, *UNESCO-IHE Institute for Water Education*, Delft, The Netherlands, ISBN: 978-0-415-56598-1 (Taylor & Francis Group).

Siebenmorgen, C.B., Sheshukov, A.Y., Douglas-Mankin, K.R., 2010. Impacts of Climate Change on Hydrologic Indices in a Northeast Kansas Watershed. Watershed Management to Improve Water Quality Proceedings, 14-17 November 2010, Hyatt Regency Baltimore on the Inner Harbor, Baltimore, Maryland.

Singh, V.P., 1995. Watershed modelling. In: V.P. Singh (ed.), Computer Models of Watershed Hydrology. *Water Resources Publication, Highlands Ranch*, CO, USA, pp. 1-22.

Sivapalan, M., Takeuchi, K. and Franks, S.W., 2003. IAHS Decade on Predictions in Ungauged Basins (PUB), 2003-2012: Shaping an exciting future for the hydrological sciences. *Hydrol. Sci. J.*, 48(6): 857-880.

Sivapalan, M., 2009. The secret to "doing better hydrological science": change the question! *Hydrol. Process.*, 23: 1391–1396. Sklash, M.G. and Farvolden, R.N., 1979. The role of groundwater in storm runoff. *J. Hydrol.*, 43, 45–65.

Snyder F.F., 1938. Synthetic unit hydrographs. *Trans Am Geophysics Union*, 19: 447-454.

Spieksma, J.F.M., 1999. Changes in the Discharge Pattern of a Cutover Raised bog During Rewetting. *Hdrol. Process.*, 13, 1233-1246.

Solomatine, D.P., 2011. Hydrological Modelling. *Treatise on Water Science*, 2: 435–457.

Soulsby, C., Neal, C., Laudon, H., Burns, D.A., Merot, P., Bonell, M., Dunn, S.M., Tetzlaff, D., 2008. Catchment data for process conceptualization: simply not enough? *Hydrological Processes*, 22: 2057–2061.

Sugawara, M., 1967. The flood forecasting by a series storage type model, *Proc. of International Symposium on floods and their computation*, Leningrad, USSR, *IAHS Publication*, 85: 1-6.

Sugawara, M., 1995. Tank Model. In: V.P. Singh (ed.), Computer Models of Watershed Hydrology. *Water Resources Publication*, Highlands Ranch, CO, USA, pp. 165-214.

Thompson, D.B., 2006. The Rational Method. *Civil Engineering Department, Texas Tech University*, Texas, USA.

Todini, E., 1988. Rainfall-runoff modelling ? past, present and future. *Journal of Hydrology*, 100(1-3): 341-352.

Twagiramungu, F., 2006. Environmental Profile of Rwanda. *Report for the National Authorising Officer of FED and the European Commission*, Kigali, Rwanda.

Twahirwa, A., 2013. Streamflow trends and climate linkages in the Rwandan catchments. *MSc thesis, National University of Rwanda*, Department of Civil Engineering, Butare, Rwanda, 76 pp.Uchida, T., Tromp-van Meerveld, I., and McDonnell, J.J., 2005. The role of lateral pipe flow in hillslope runoff response: An intercomparison of nonlinear hillslope response. *J. Hydrol.*, 311: 117–133.

Uhlenbrook S., 2006. Catchment hydrology — a science in which all processes are preferential, *Hydrological Processes*, HPToday, 20, 16, 3581–3585, DOI: 10.1002/hyp.6564.

Uhlenbrook S., 2007. Biofuel and water cycle dynamics: what are the related challenges for hydrological processes research, *Hydrological Processes*, Volume 21, Issue 26, 3647-3650, doi: 10.1002/hyp.6901.

Uhlenbrook S. 2009. Climate and man-made changes and their impacts on catchments. In Water Policy 2009, Water as Vulnerable and Exhaustible Resources, Proceedings of the Joint Conference of APLU and ICA, 23–26 June 2009, Kovar P., Maca P., Redinova J. (eds). Prague: Czech Republic; page 81–87.

Uhlenbrook, S. and Hoeg S., 2003. Quantifying uncertainties in tracer-based hydrograph separations - A case study for two, three and five component hydrograph separations in a mountainous catchment. *Hydrological Processes*, 17(2), 431-453.

Uhlenbrook, S. and Leibundgut, Ch., 2002. Process-oriented catchment modelling and multiple response validation. *Hydrological Processes*, 16, 423-440.

Uhlenbrook, S., Didszun, J. and Wenninger, J., 2008. Sources areas and mixing of runoff components at the hillslope scale – a multi-technical approach. *Hydrological Sciences Journal*, 53(4).

Uhlenbrook, S., Frey, M., Leibundgut, C., and Maloszewski, P., 2002. Hydrograph separations in a mesoscale mountainous basin at event and seasonal timescales, *Water Resour. Res.*, 38(6), 1096–1110, doi:10.1029/2001wr000938.

Uhlenbrook, S., Seibert, J., Leibundgut, C. and Rodhe, A., 1999. Prediction uncertainty of conceptual rainfall-runoff models caused by problems to identify model parameters and structure. *Hydrological Sciences Journal*, 44(5), 779-799.

UNEP, 2005. Connecting poverty and ecosystem services: A series of seven country scoping studies, focus on Rwanda. *United Nations Environmental Programme (UNEP) and the International Institute for Sustainable Development (IISD)*, Nairobi, Kenya.

UR Research Agenda, 2007. Water Resources and Environmental Management Project, Research Agenda for 2007-2011 and beyond. *National University of Rwanda*, Faculty of applied Sciences, Butare, Rwanda.

USACE, 2008. Technical Workshop on Watershed Modelling with HEC-HMS. U.S. Army Corps of Engineers, California Water and Environmental Modelling Forum, Sacramento, California.

USACE, 2010. HEC-GeoHMS Geospatial Hydrologic Modelling Extension, v5.0. User's Manual. US Army Corps of Engineers, Hydrologic Engineering Center, October 2010.

USACE-HEC, 2003. Geospatial Hydrologic Modelling Extension, HEC-GeoHMS, v1.1. User's Manual. US Army Corps of Engineers, Hydrologic Engineering Center, December 2003.

USACE-HEC, 2006. Hydrologic Modelling System, HEC-HMS, v3.0.1. User's Manual. *US Army Corps of Engineers, Hydrologic Engineering Center*, April 2006.

USACE-HEC, 2010. Hydrologic Modelling System, HEC-HMS, v3.5. User's Manual. *US Army Corps of Engineers, Hydrologic Engineering Center*, August 2010.

USEPA, 2012. Safe and Sustainable Water Resources, Strategic Research Action Plan 2012-2016. *United States Environmental Protection Agency.*

Uvin, P., 1998. Aiding violence, the development enterprise in Rwanda. *Kumarian Press.,* West Hartford, CT.

van Breukelen, B.M., Groen, M.M.A., Groen, J., van Huissteden, J., de Jeu, R.A.M., Post, V.E.A., Schellekens, J., Smit, P. and Waterloo, M.M., 2008. Handbook for Field Hydrological Measurements, Field course Netherlands Reader, Faculty of Earth and Life Sciences, VU University Amsterdam.

van den Berg, H.W. and Bolt, R.H., 2010. Catchment analysis in the Migina marshlands, southern Rwanda. MSc thesis, *Vriije University (VU) Amsterdam*, The Netherlands, 120p.

Van Griensvena, A., Xuan, Y., Haguma, D. and Niyonzima, W., 2008. Understanding riverine wetland-catchment processes using remote sensing data and modelling. 4th Biennial Meeting of iEMSs, http://www.iemss.org/iemss2008/index.php?n=Main.Proceedings.Verdoodt, A. and van Ranst E., 2003. Land Evaluation for Agricultural Production in the Tropics: A Large-Scale Land Suitability Classification for Rwanda, *Laboratory of Soil Science*, Ghent University.

Verdoodt, A. and van Ranst, E., 2006. Soil information system of Rwanda, a usefull tool to indentify guidelines towards sustainable land management". *Laboratory of Soil Science, Ghent University*, Afrika Focus, 19: nr.1-2.Viessman, J.W., Lewis, L.G. and Knapp, W.J., 1989. Introduction to Hydrology, 3rd *ed., New York*, 780 p.

Wagener, T., Sivapalan, M., Troch, P. and Woods, R., 2007. Catchment Classification and Hydrologic Similarity. *Geography Compass*, 1(4): 901– 931.Wagener, T., Sivapalan, M. and McGlynn, B., 2008. Catchment Classification and Services-Toward a New Paradigm for Catchment Hydrology Driven by Societal Needs, *Encyclopedia of Hydrological Sciences*, has 320, Edited by M G Anderson.

Waterloo, M.J., Post, V.E.A., and Horner, K., 2007. Introduction to Hydrology, Lecture notes, Course Code 450024, 5 ECTS, *Vrije University Amsterdam*, The Netherlands.

WaterNet, 2008. Water and Sustainable Development for Improved Livelihoods. WaterNet/WARFSA/GWP-SA, *International conference on 9th WaterNet Symposium*, 29-31 October 2008, Johannesburg, South Africa.

Wels, C., Cornett, R. J., and Lazerte, B.D., 1991. Hydrograph separation: a comparison of geochemical and isotopic tracers. *Journal of Hydrology*, 122 (1–4), 253–274.

Wenninger, J., Uhlenbrook, S., Lorentz, S. and Leibundgut, C., 2008. Identification of runoff generation processes using combined hydrometric, tracer and geophysical methods in a headwater catchment in South Africa. *Journal of Hydrological Sciences*, 53(1), 65-80.

Western, A.W., Grayson, R.B., Bl¨oschl, G., Willgoose, G.R., McMahon, T.A., 1999. Observed spatial organization of soil moisture and its relation to terrain indices. *Water Resources Research*, 35, –.

WMO, 2008. Guide to Hydrological Practices. *Volume-From measurements to hydrological Information*, WMO-No. 168, World Meteorological Organization, Geneva.

WMO, 2010. Manual on stream gauging. *Volume II, Computation of Discharge*, WMO-No. 1044, World Meteorological Organization, Geneva.

Word Bank, 2006. Water Resources and Freshwater Ecosystems in Rwanda. *World Resources Institute*, visited in October 2008, http://earthtrends.wri.org

World Bank (2008), Second Rural Sector Support Project to Support Intensification of Agricultural Production Systems and Commercialization of Agricultural Products. *World Bank Press Release*, No: 2008/407/AFR.

WRPM, 2006. Water Resources Planning and Management Project (WRPM). http://wrpmp.nilebasin.org/index.php?Itemid=37&id=27&option=com_content&task=vie.

Yawson, D.K., Kongo, V.M. and Kachroo, R.K., 2005. Application of linear and nonlinear techniques in river flow forecasting in the Kilombero River basin, Tanzania. *Hydrological Sciences Journal*, 50(5): 783-796.

Yener, M.K., Şorman, A.Ü., Şorman, A.A., Şensoy, A. and Gezgin, T., 2007. Modelling studies with HEC-HMS and runoff scenarios in Yuvacik basin, Turkey. *International Congress on river basin management*, 4: 621-634.

Zhang, Q., Singh, P.V., Sun, P., Chen, X., Zhang, Z. and Li, J., 2011. Precipitation and streamflow changes in China: Changing patterns, causes and implications. *Journal of Hydrology*, 410(3-4): 204-216.

Zhang, W., Yan, Y., Zheng, J., Li, L., Dong, X. and Cai, H., 2009. Temporal and spatial variability of annual extreme water level in the Pearl River Delta region, China. Global and Planetary Change 69 (2009) 35–47. Zhang, X., Zhang, L., Zhao, J., Rustomji, P. and Hairsine, P., 2008. Responses of streamflow to changes in climate and land use/cover in the Loess Plateau, China. *Water Resour. Res.*, 44 W00A07, DOI:10.1029/2007WR006711.

PUBLICATIONS BY THE AUTHOR

Peer-reviewed publications

1. Munyaneza, O., Nzeyimana, Y.K. and Wali, U.G., 2013. *Hydraulic structures design for flood control in the Nyabugogo wetland, Rwanda*. Nile Water Sci. Eng. J., 6(2): 26-37.

2. Munyaneza, O., Nzeyimana, G., Nsengimana, H., Uzayisenga, Chr., Uwimpuhwe, Ch. and Nduwayezu, J.B., 2012. *Surface Water Resources Assessment in the Rwasave Marshland, southern Rwanda*. Nile Water Sci. Eng. J., 5(2): 58-70.

3. Munyaneza, O., Wenninger, J., and Uhlenbrook, S., 2012. *Identification of runoff generation processes using hydrometric and tracer methods in a meso-scale catchment in Rwanda*. Hydrol. Earth Syst. Sci., 16: 1991–2004.

4. Munyaneza, O., Wali, U.G.; Ufiteyezu, F. and Uhlenbrook, S., 2011. *A simple Method to Predict River Flows in the Agricultural Migina Catchment in Rwanda*. Nile Water Sci. Eng. J., 4(2): 24-36.

5. Munyaneza, O., Ndayisaba, C., Wali, U.G., Mulungu, M.M.D., and Dulo, O.S., 2011. *Integrated Flood and Drought Management for Sustainable Development in the Kagera River Basin*. Nile Water Sci. Eng. J., 4(1): 60-70.

6. Munyaneza, O., Uhlenbrook S., Wenninger, J., van den Berg, H., Bolt H. Rutger, Wali G.U. and Maskey S., 2010. *Setup of a Hydrological Instrumentation Network in a Meso-Scale Catchment-the case of the Migina Catchment, Southern Rwanda*. Nile Water Sci. Eng. J., 3(1): 61-70.

7. Munyaneza, O., Wali, U.G, Uhlenbrook, S., Maskey, S. and Mlotha, M.J, 2009. *Water Level Monitoring using Radar Remote Sensing Data: Application to Lake Kivu, Central Africa*. Physics and Chemistry of the Earth, 34: 722-728.

Publications accepted and submitted

1. Munyaneza, O., Mukubwa, A., Maskey, S., Wenninger, J. and Uhlenbrook, S., 2013. *Assessment of surface water resources availability using catchment modelling and the results of tracer studies in the meso-scale Migina Catchment, Rwanda*. Hydrol. Earth Syst. Sci. Discuss., 10: 15375-15408.

2. Munyaneza, O., Maskey, S., Uhlenbrook, S., Twahirwa A. and Wenninger, 2012. *Streamflow trends and climate linkages in meso-scale catchments in Rwanda*. IWA Publishing, submitted.

Conference presentations

1. Munyaneza, O., Nzeyimana, Y.K., Wali, U.G., 2013. *Hydraulic structures design for flood control in the Nyabugogo wetland, Rwanda*. 3rd International Young Water Professionals Conference on *"Securing our water and energy resources in the face of climate change"*, Nairobi, Kenya, 9th–11th December 2013, Oral presentation.

2. Munyaneza, O., 2013. *Assessment of Surface Water Resources Availability in the Agricultural Rwasave Marshland - contribution to Food Security in Rwanda*. World Water Day, 22 March 2013, Butare, Rwanda, Oral presentation.

3. Munyaneza, O., Maskey, S., Uhlenbrook, S., Twahirwa, A. and Wenninger, J., 2012. *Streamflow trends and climate linkages in meso-scale catchments in Rwanda*. Proceedings of the Climate Change, 2nd International Young Water Professionals Conference on *"Water for the future: The contribution of the Youth"*, Kigali Serena Hotel, Rwanda, 9th–12th December 2012, best oral paper presentation Award.

4. Munyaneza, O., Nzeyimana, G., Nsengimana, H., Uzayisenga, Chr. and Uwimpuhwe, Ch., 2012. *Surface Water Resources Assessment in Rwasave Marshland - contribution to poverty reduction in Rwanda*. Proceedings of the Hydrology and Water Resources, p67, VII International Conference

on Environmental Hydrology, Concorde El Salam Hotel Cairo-Egypt, 25-27 Sept 2012. Best oral paper presentation Award.

5. Munyaneza, O. and Maniraruta, G., 2011. *Surface Water Resources Availability in the Migina Sub-catchment: Contribution to Food Security in Rwanda.* Proceedings of the Water Management, Interdisciplinary Ph.D. Conference in Sustainable Development, Columbia University, New York, USA, May [6th]-[7th], 2011. Oral presentation.

6. Munyaneza, O. Wali, U. G., Ufiteyezu, F. and Uhlenbrook, S., 2011. *River Flow Quantification in an Agricultural Dominated Catchment: A Case study of Migina, Rwanda.* Proceedings, International Hydrological Programme (IHP) Conference, Rwanda National Commission for UNESCO, Umubano Hotel, Kigali, Rwanda, 28-29 April 2011. p17-26, Oral presentation.

7. Munyaneza, O., Ndayisaba, C., Wali, U.G., Bizimana, R., Uzayisenga, Ch., and Nkurunziza, T., 2010. *Identification and quantification of runoff generation processes during floods and droughts in the Migina catchment, Southern Rwanda.* International NBCBN Conference, Cairo, Egypt, 28-30 November 2010. Oral and Poster presentation.

8. Munyaneza, O., Dulo, S.O., Odira, P.M.A., Nyabenze, M., Shaka, A., Nyadawa, M., Mulungu, D.M., Ndayisaba, C., Mwogeza, M., Mwandoe, E., Sousa, H.K. and Fadul, H., 2010. *Integrated Flood and Drought Management for Sustainable Development in the Nile Basin: The case of Nzoia and Kagera River Basins.* International NBCBN Conference, Cairo, Egypt, 28-30 November 2010. Oral and Poster presentation.

9. Munyaneza, O., Uhlenbrook, S. and Wenninger, J., 2010. *Identification of runoff generation processes during floods and low flows using hydrometric data and tracer methods in the Migina catchment in Rwanda.* Proceedings of the Hydrology Cycle, p. 65, European Geosciences Union (EGU) International Leonardo Tropical Conference, Belvaux, Luxembourg, 10-12 Nov. 2010, Poster presentation.

10. Munyaneza, O., Uhlenbrook, S., Maskey, S., Wali, U. G. and Wenninger, J., 2009. *Hydrological and climatic data availability and preliminary analysis in Rwanda.* Proceedings of the Hydrology, 10th International WaterNet/WARFSA/GWP-SA Symposium, Entebbe, Uganda, 28-30 October 2009. Poster presentation.

11. Munyaneza, O., Musoni, J.P. and Wali, U.G., 2009, *Runoff coefficient classification on Nyabugogo catchment.* Proceedings of the Hydrology, 10th International WaterNet/WARFSA/GWP-SA Symposium, Entebbe, Uganda, 28-30 October 2009. Oral presentation.

12. Munyaneza, O., Uhlenbrook, S., Maskey, S., Wali, U.G. and Wenninger, J., 2009. *Hydro-climatic instrumental set-up in Migina catchment for sustainable water resources planning and management in Rwanda.* Proceedings of the Flood, Sedimentation and Erosion, Regional NBCBN Conference, Kigali, Rwanda, 01-03 December 2009. Oral presentation.

13. Munyaneza, O., Uhlenbrook, S., Wali, U.G., Wenninger, J. and Maskey, S., 2009. *Hydrological Instrumentation of Meso-Scale Catchment: A case study of Migina Catchment, Southern Rwanda.* Proceedings of the Science, Engineering and Technology, 2nd International UR Scientific Research Conference, Butare, Rwanda, 19-21 October 2009. Oral presentation.

14. Munyaneza, O., Wali, U.G., Mlotha, M.J., Lasry, F. and Uhlenbrook, S., 2008. *Water level Monitoring using Radar Remote Sensing data. Application on Lake Kivu.* Proceedings of the Hydrology, 9th International WaterNet/WARFSA/GWP-SA Symposium, Johannesburg, South Africa, 29-31 October 2008. Oral presentation.

ABOUT THE AUTHOR

Eng. Omar Munyaneza was born on 30th September 1976 in Umutara, Gatsibo District, Eastern Province of Rwanda. He obtained his Bachelors Degree (B.Sc.) in Civil Engineering in October 2004 from National University of Rwanda (NUR). In 2007, Omar graduated with a Master of Science (MSc) at the programme called Water Resources and Environmental Management (WREM) in collaboration with UNESCO-IHE Institute for Water Education, Delft, the Netherlands.

Omar worked as Technical Director in a private Civil Engineering Company (ECOSEKAT) at Kigali from November 2004 to September 2005. In October 2005, he joined Institute of Scientific and Technological Research (IRST) in Butare as Research Assistant, then as Director of Rusizi Research Station in Western Province of Rwanda from May 2007 till February 2008. In March 2008, Omar joined National University of Rwanda (NUR), Department of Civil Engineering as an Assistant Lecturer and he was promoted to the rank of Lecturer in November 2011.

In July 2008, Omar started a PhD study at UNESCO-IHE Institute for Water Education, Delft, The Netherlands. He carried out field experiments in Rwanda, southern Province where he developed and maintained monitoring of a hydrometeorological instrumentation network. This includes 13 rain gauges and 3 tipping buckets gauges, 2 evaporation pans, 1 weather station, 5 river gauging stations and 11 shallow piezometers for groundwater monitoring.

During his PhD research, Omar followed the Educational Programme of SENSE (Socio-Economic and Natural Sciences of the Environment) and obtained a Certificate in December 2013. He presented his work at several international conferences, where he got two Awards for best oral presentation. One in the International Conference on *"Environmental Hydrology"* held at Cairo, Egypt on 25th–27th September 2012, and another Award during the 2nd International Young Water Professionals Conference on *"Water for the future: The contribution of the Youth"* held at Kigali, Rwanda on 9th–12th December 2012. Omar was also actively involved in BSc and MSc courses and supervised several students. For the courses of "Hydrology", "Fluid Mechanics", "Research Methodology" and "Water Resources and Drainage Systems", he wrote lecture notes, gave lectures, and guided students for field work in the Migina catchment, southern Rwanda.

Since 2010 Omar is a member of World Meteorological Organization (WMO)/Associated Programme on Flood Management (APFM), a member of International Hydrological Programme (IHP) for UNESCO, and he is a Coordinator of Nile Basin Capacity Building Network (NBCBN) under Rwanda Flood Management Cluster. Since April 2013, Omar is a Rwanda National Training Coordinator of Global Water Partnership (GWP) under the Programme of WACDEP Capacity Development in Africa (WACDEP: Water, Climate and Development Programme).

Omar is now a lecturer at University of Rwanda (UR) and Coordinator of NICHE Project (Netherlands Initiative for Capacity development in Higher Education) and also a visiting lecturer of Engineering Hydrology and Fluid Mechanics in Civil Engineering Department at Saint Joseph Integrated Technological College (ITC) at Kigali-Rwanda since 2012. Along with these responsibilities, Omar finalysed his PhD thesis in August 2013.

Omar is married to Eng. Christine and they have a daughter Leïla Munezero (7 years) and a son Laïq Munyaneza (3 years).

The findings of his PhD resulted in a number of publications in peer-reviewed journals and conference proceedings.

The SENSE Research School Certificate

Netherlands Research School for the
Socio-Economic and Natural Sciences of the Environment

C E R T I F I C A T E

The Netherlands Research School for the
Socio-Economic and Natural Sciences of the Environment
(SENSE), declares that

Omar Munyaneza

born on 30 September 1976 in Umutara, Gatsibo, Rwanda

has successfully fulfilled all requirements of the
Educational Programme of SENSE.

Delft, 18 December 2013

the Chairman of the SENSE board

Prof. dr. Rik Leemans

the SENSE Director of Education

Dr. Ad van Dommelen

The SENSE Research School has been accredited by the Royal Netherlands Academy of Arts and Sciences (KNAW)

K O N I N K L I J K E   N E D E R L A N D S E
A K A D E M I E   V A N   W E T E N S C H A P P E N

T - #0435 - 101024 - C126 - 240/170/7 - PB - 9781138026575 - Gloss Lamination